Probability and Statistics in Civil Engineering

An Introduction

PROBABILITY AND STATISTICS IN CIVIL ENGINEERING

An Introduction

G. N. Smith MSc, PhD, CEng, MICE

Civil Engineering Department
Heriot-Watt University

Nichols Publishing Company
New York

Nichols Publishing Company (W. G. Nichols, Inc.)
PO Box 96, New York, New York 10024 and
155 West 72nd Street, New York, New York 10023.

Library of Congress Cataloging-in-Publication Data

Smith, G. N. (Geoffrey Nesbitt)
  Probability and statistics in civil engineering.

  1. Structural engineering—Statistical methods.
2. Soil mechanics—Statistical methods.
3. Probabilities.   I. Title.
TA640.2.S55 1986   624'.01'5192   86-8344

First published in the United States of America
by Nichols Publishing Company 1986
First published in Great Britain by
Collins Professional and Technical Books 1986
8 Grafton Street, London W1X 3LA

Filmset by Eta Services (Typesetters) Ltd.,
Beccles, Suffolk
Printed and bound in Great Britain by
Mackays of Chatham, Kent

Contents

7. Matrix Algebra   159

Matrices, Elements, Square matrix, Leading diagonal, Trace, Diagonal matrix, Unit Matrix, Triangular matrix, Column matrix, Addition, subtraction and multiplication of matrices, Transpose, Symmetric matrix, Null matrix, Multiplication of vectors, Determinant, Elimination methods, Inverse matrix, Singular matrix, Cofactor matrix, Adjoint matrix, Eigenvalues and eigenvectors, Spectral Matrix, Transformation methods, Jacobi's method.

8. Correlated and Non-normal Variables   183

Multivariate distributions, Scatter diagrams, Regression lines, Joint probability functions, Joint pmf, Joint pdf, Covariance, Linear correlation coefficients, Standard error of the estimate, Correlated variables, Covariance matrix, Variables with non-normal distributions, Load combination.

9. The Reliability of Geotechnical Structures   209

Site investigations, Soil sampling theory, Probabilistic treatment of the substrata, Spatial uniformity, Directional trends, Retaining wall analysis.

Preface

Until recently the ability of a structural element to withstand a particular loading, or to not deflect more than a prescribed limit, was expressed in terms of a single number, the factor of safety, F.

A drawback of this method is that the value of the factor of safety is simply a number obtained from a deterministic approach in which there is no allowance for any inherent variability within the design parameters and it is, perhaps, not surprising that there have been instances of structural failure where the calculated factor of safety was actually greater than 1.0.

Lumb (1970), speaking in the context of geotechnical engineering, summed up the situation:

'The traditional safety factor concept has the serious disadvantage that the actual variability of the soil strength is not directly taken into account and, consequently, a particular conventional safety factor value does not necessarily have the same meaning for all soils. Comparison of different designs with different soil types, or even different designs with the same soil type, is not easy, unless the conventional safety factors are so large as to preclude any practical risk of failure.'

The factor of safety, far from being of constant value, is really a random variable whose variability is due to the variability of the applied loads and the strength parameters of the structure.

If failure is defined as the event of F achieving a value equal to or less than 1.0 then the probability of this event is the probability of failure, P_f.

In Britain the first major step to allow for civil engineering uncertainties in design took place in 1972 when the Code of Practice CP110, *The Structural Use of Concrete*, was published by the British Standards Institution. This code adopted the policy of limit state design and probability theory was used, albeit indirectly, by the introduction of characteristic values.

Since 1972 the pressure for change has not diminished. Most of the proposed Eurocodes for structural design are now in draft form and advocate limit state design. There is a strong chance that these codes will either list values of partial safety factors, determined with the help of probability theory, or will make use of the reliability index which is becoming recognised as a powerful alternative to the use of partial safety factors in civil and structural engineering design. Along with this change there is a need for consultants, students, lecturers and research workers in civil engineering to at least become familiar with these developments and the new terms they involve.

This book is intended to present a straightforward summary of the most important aspects of statistics and probability theory that are relevant in civil engineering. Within these limitations the text is complete in that it should be possible for the reader to work through it without reference to other books.

The first three chapters deal with the fundamentals of statistics and its application to probability theory. At the end of these chapters there are exercises which, it is hoped, will be of assistance in the understanding of the subject matter. Those with knowledge of this material will be able to commence reading the book at chapter four, where the principles of reliability analysis are first discussed.

The author would like to take this opportunity of thanking those colleagues who gave helpful suggestions and encouragement during the preparation of this book. In particular he would like to thank Dr M. A. Paul of the Civil Engineering Department of Heriot-Watt University and Dr R. T. Murray of the Road Research Laboratory at Crowthorne.

G. N. Smith

Chapter One

Basic Probability Theory

Sets and events

The study of events and the probability of their happenings inevitably draws one towards the idea of the set.

In a test series of measurements the mean value obtained is an event resulting from the whole set of measured values.

A set is therefore a collection of items and, as with an event, is usually designated by a capital letter, A, B, C, etc. The individual elements that make up a set are generally denoted by lower case letters, a, b, c, ... For example, for set A:

$$A = a_1, a_2, a_3, a_4$$

Since the arrangement of elements does not affect a set, A is also given by:

$$A = a_3, a_4, a_1, a_2$$

The convention $a_1 \in A$ simply means that a_1 is an element of the set A.

In most civil engineering situations a set is defined by the listing of the elements within it, such as the measurements obtained for a particular test. However there are often occasions when it is not possible to determine the total elements of a set, although we know they exist, such as the infinite set of soil samples that could be collected from a particular stratum.

In such a situation, although the full set cannot be listed, the properties of the set can. For example a set B, consisting of all even numbers between 2 and 100, could be specified as:

$$B = [b \, ; \, b \text{ is an even number between 2 and } 100]$$

where ';' means 'given that' or 'such that'.

Obviously set B could also have been listed as:

$$B = [2, 4, 6, 8, \ldots, 98, 100]$$

The universal set

The complete collection of all possible elements of a set is known as the universal set or the sample space and given the symbol Ω, the Greek letter omega, or the capital letter S.

Figure 1.1 shows the sample space, i.e. all the possible events, involved in the scores obtained from the throwing of two dice.

The sample space, such as the total 36 elements of Fig. 1.1, represents the certain event, in this case the event 'there will be some score'.

An impossible event is one which is outside the sample space, such as the event (7, 1) in Fig. 1.1.

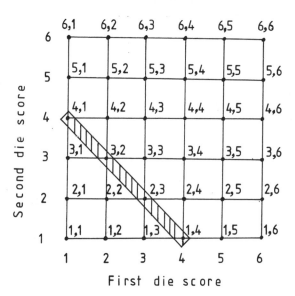

Fig. 1.1   Sample space for score of two dice

The subset

If B is a set of elements taken from a universal set, A, then B is referred to as a subset of A. This is expressed as:

$$B \subset A \quad \text{or} \quad A \supset B$$

meaning 'B is contained in A' or 'A contains B' respectively.

In Fig. 1.1 the subset [(4, 1), (3, 2), (2, 3), (1, 4)] represents the total number of ways of obtaining the score '5'. The event 'scoring 5' is by no means a single event as it can occur in 4 different ways.

Simple event

An event that can only occur once, such as the scoring of double 1, is referred to as a simple, or elementary, event.

Compound event

An event that can occur in more than one way is a compound event.

Union of sets (A ∪ B)

The union of two sets, A and B, is the set which contains all the elements that are in either A or B.

Intersection of sets (A ∩ B)

The intersection of two sets, A and B, is the set which contains all the elements that are in both A and B.

Example 1.1

The values of the blow count, N, measured during a series of penetration tests on a sand deposit were found to range between 1 and 15. The sample space representing all possible N values, therefore, is a set of figures, 1, 2, ..., 14, 15 and is shown in Fig. 1.2.

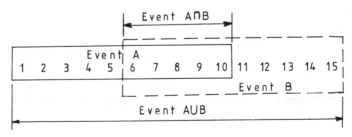

Fig. 1.2   Union and intersection of sets

If event A is that $1 \leqslant N \leqslant 10$ and the event B is that $6 \leqslant N \leqslant 15$ then the union of A and B, (A ∪ B), is the event that $1 \leqslant N \leqslant 15$ and the intersection of A and B, (A ∩ B) is the event that $6 \leqslant N \leqslant 10$.

The complement of a set

If *B* is a subset of A then A > B and the set (A − B) is called the complement of B relative to A and given the symbol $\bar{B}_A$.

If S is the total sample space then the set (S − B) is known as the complement of B and given the symbol $\bar{B}$.

The complement of A ∪ B is written as $\overline{A \cup B}$.

Difference between sets

The set containing those elements of A that are not in B is the set $(A \cap \bar{B})$. Such a set is often referred to as the difference between A and B, as $(A \cap \bar{B})$ is numerically equal to (A − B).

Example 1.2

A total sample space is the set (1, 2, 3, 4, 5, 6, 7, 8, 9, 10).

If A is (1, 2, 4, 5, 8, 9) and B is (3, 5, 6, 7, 8) show numerically that $(A \cap \bar{B})$ equals (A − B).

Solution

$$\bar{B} = 1, 2, 4, 9, 10$$

Hence:

$$(A \cap \bar{B}) = (1, 2, 4, 9)$$

and:

$$(A - B) = (1, 2, 4, 9)$$

Note: The connotation (A − B) has a different meaning in vectorial algebra where (A − B) would be equal to (1, 2, −3, 4, −6, −7, 9).

The Venn diagram

A universe, or a sample space, S, and its subsets can be presented in a pictorial form by using these diagrams.

The universal set, S, is represented as a rectangle with its subsets lying within it, as seen in Fig. 1.3a. The shaded area of Fig. 1.3b illustrates the difference set (A − B) and the shaded area of Fig. 1.3c represents $\bar{A}$, the complement of set A.

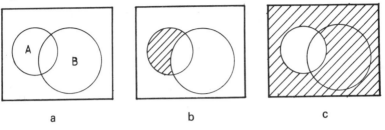

Fig. 1.3 The Venn diagram

The algebra of sets and events

Figure 1.1 illustrates that if an event can happen in several different ways then the event is a subset of the set of total possible events. It can be seen, therefore, that the algebra applicable to sets is identical to that for events. If A and B are events then, in set theory, the symbols mean:

(1) $A \cup B$ = the event 'the happening of either A or B or both'
(2) $A \cap B$ = the event 'the happening of both A and B'
(3) $\bar{A}$ = the event 'the non-happening of A'
(4) $A \cap \bar{B}$ = the event 'the happening of A but not B' = $(A - B)$

The Venn diagrams of Fig. 1.4 illustrate various set operations.

The most important theorems of set algebra are set out below and can be demonstrated by a study of the appropriate Venn diagrams.

Commutative law: $A \cup B = B \cup A$
                   $A \cap B = B \cap A$

Associative law: $A \cup (B \cup C) = (A \cup B) \cup C = A \cup B \cup C$
                   $A \cap (B \cap C) = (A \cap B) \cap C = A \cap B \cap C$

Distributive law: $A \cap (B \cup C) = (A \cap B) \cup (A \cap C)$
                   $A \cup (B \cap C) = (A \cup B) \cap (A \cup C)$

Complementary laws: $A - B = A \cap \bar{B}$

De Morgan's laws: $\bar{A} \cup \bar{B} = \overline{A \cap B}$
                   $\bar{A} \cap \bar{B} = \overline{A \cup B}$

Example 1.3

Illustrate the distributive law, $A \cap (B \cup C) = (A \cap B) \cup (A \cap C)$, by means of the Venn diagram.

Description	Mathematical expression	Venn diagram
<u>Mutually exclusive events</u> A and B (no common elements)	$A \cap B = 0$	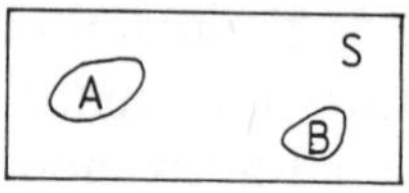 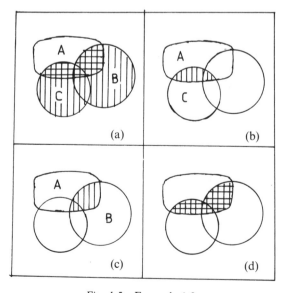
<u>B is a subset of A</u> (all elements of B are included in A)	$B \subset A$	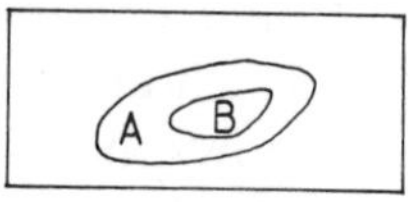
<u>Union of A and B</u> (all elements that are in either A or B)	$A \cup B$	
<u>Intersection of A and B</u> (all elements in both A and B)	$A \cap B$	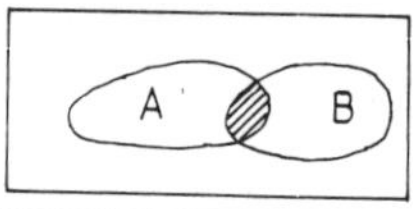
<u>Difference between A and B</u> (elements in A but not in B)	$A - B$	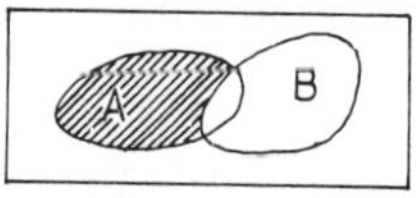
<u>Complementary set $\bar{A}$</u> (elements not in A)	$\bar{A} = S - A$	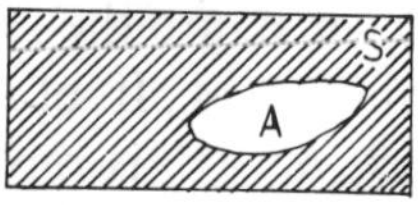

Fig. 1.4 Set operations

(a)   (b)

(c)   (d)

Fig. 1.5 Example 1.3

Solution

The three sets A, B and C are shown in Fig. 1.5.

In Fig. 1.5a the union of B and C, i.e. (B ∪ C) is shown with vertical lines whilst the intersection of (B ∪ C) with A, i.e. A ∩ (B ∪ C), is marked with additional horizontal lines.

Figures 1.5b and 1.5c respectively show the intersections of A and C and A and B, which are both marked by vertical lines.

Figure 1.5d shows the union of (A ∩ B) and (A ∩ C), which is marked with both horizontal and vertical lines, and is seen to be identical to Fig. 1.5a, thus proving the theorem.

Note: The reader can prove the theorem numerically by assuming sets of values for each of the three sets A, B and C.

Example 1.4

By means of Venn diagrams prove the theorem A − B = A ∩ B̄.

Solution

When stated in words the theorem is, 'The elements contained in a set A, but not in a set B, are the same elements common to both set A and the complement of set B'. If set A and set B are as shown in Fig. 1.6a then the difference set, A − B, is represented by the hatched area shown.

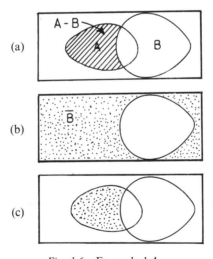

Fig. 1.6 Example 1.4

The dotted area of Fig. 1.6b represents the complement set, B̄, and it is fairly obvious that the dotted area of Fig. 1.6c, which represents the elements common to A and $\bar{B}$, is the same as the hatched area of Fig. 1.6a. Hence:

$$A - B = A \cap \bar{B}$$

Note: The above axiom can be illustrated by considering the elements within the sets:

$$A - B = [x ; x \in A \text{ and } x \notin B]$$
$$= [x ; x \in A \text{ and } x \in \bar{B}] = A \cap \bar{B}$$

Probability

The probability that a particular event, A, will happen is expressed mathematically as P[A].

If the event A will never happen, e.g. pigs will fly, then the value of P[A] will be 0 whereas if event A will happen, e.g. the world will end sometime, then P[A] is 1.

Probability values are classified in one of two ways, depending upon how they are estimated, as follows.

PRIOR PROBABILITY VALUES

Prior probability values are obtained by the subjective, or degree of believe, point of view which involves predictions based on past experience and expertise (i.e. a priori judgement) of the decision maker.

The most commonly quoted source of prior probability values is the throwing of dice. With a single die there are six possible scores and, if the die is fair, each of them is equally likely to be obtained at any one throw. The probability of obtaining a particular score is therefore one in six, expressed as $\frac{1}{6}$.

Example 1.5: Prior probability

Determine the probability of drawing an ace from a full pack of cards.

Solution
There are 4 aces in a pack and a total of 52 cards. Hence $n = 4$ and $N = 52$. *Therefore, the probability of drawing an ace $= n/N = \frac{4}{52} = \frac{1}{13}$.*

Note: Most civil engineers have little difficulty in accepting this sort of reasoning but many encounter difficulties when extending the idea of degree of belief to civil engineering situations. For instance, few would be willing to accept that the probability of a rock fault existing at some site is 60%. Most would argue that, as the fault either exists or does not exist, then the probability is either 1 or 0.

It is in these situations that Bayes' theorem, described later in this chapter, can be of assistance in decision making.

POSTERIOR PROBABILITY VALUES

Posterior probabilities are estimated with hindsight, i.e. by the use of a frequenistic approach involving predictions based on a study of a series of repeatable events or tests.

Example 1.6: Posterior probability

Forty-five control tests were carried out on a long stretch of compacted subgrade. Five tests yielded results that were below specification.

If a further 10 tests had been carried out how many of these tests could have been expected to have given results below specification?

Solution
Probability of test results below specification is $\frac{5}{45} = \frac{1}{9}$.

For a further set of ten tests the expected number of results below specification is $\frac{10}{9} = 1.1$ i.e. one test.

Prior and posterior probabilities in civil engineering

Possibly because of their training most civil engineers tend to accept the frequenistic more easily than the subjective approach but, as most civil engineering design work involves posterior probabilities, this is generally no disadvantage. However, whilst suitable for most design situations, the frequenistic approach cannot be applied to the case of an unrepeatable event, such as the making of a design decision.

The estimation of prior probabilities of civil engineering situations, as opposed to dice throwing, can only be obtained by an a priori approach, i.e. subjective judgement of the operator involving his previous experience.

On the face of it the statements $P[A] = 1$ and $P[B] = 0$ imply absolute certainty and there are many situations, such as life and death, when this is so. However, in civil engineering, with posteriori judgement, one cannot assume absolutely that because an event happened in the past it will do so again in the future. Similarly, with the degree of belief approach, a civil engineering prior probability value can hardly be regarded as certain.

Generally speaking, when the statement $P[A] = 1$ occurs in this text it means that it is considered that A will most probably occur, not that it will occur.

Mutually exclusive events

If there is a set of events A, B, C, ... such that the happening of one excludes the happening of the others then we say that the events A, B, C, ... are mutually exclusive.

An example of mutual exclusion would be the acceptance of a tender from among several submitted. If contractor A is successful in his bid then there is no possibility of contractors B, C, etc. also being successful.

The summation law – union probability

This law applies to mutually exclusive events and states that for a series of mutually exclusive events, the union probability of at least one of these events occurring is equal to the sum of the separate probabilities of the events.

Consider three events, A, B and C. The probability that any one of these events will occur is:

$$P[A \cup B \cup C] = P[A] + P[B] + P[C]$$

(It may help readers if they consider the union symbol, $\cup$, to represent the word, 'or'.)

Example 1.7

Examples of the summation law are:

(i) The tossing of a fair coin:
     The probability of a head $= P[A] = 0.5$ or 50%
     The probability of a tail $= P[B] = 0.5$ or 50%
     Probability of either a head or a tail $= P[A \cup B] = P[A] + P[B] = 1.0$ or 100%.

(ii) A set of strength measurements of a particular material:
     $P[A] =$ the probability of the actual strength being equal to or less than the mean value $= 0.5$.
     $P[B] =$ the probability of the actual strength being equal to or greater than the mean value $= 0.5$.
     $P[A \cup B] = P[A] + P[B] =$ the probability that the actual strength

is either equal to or is greater or smaller than the mean value = 1.0.

Independent events

If we have a set of possible events such that the happening of any one event has no effect on the probabilities of the happening of the others then the events are said to be independent.

If a perfect random number generator is programmed to produce integers between 1 to 100 then the production of each number by the generator will be an independent event. This means that if the generator was to produce the number 24 in two consecutive intervals then the chance of it producing a further 24 in the next interval is exactly the same as its chance of producing any of the other numbers. However, it should be said that, if the generator did indeed produce three 24s, one after the other, there would be more than a little entitlement to consider that the generator was biased rather than perfect.

The multiplication law – joint probability

This law states that, for a series of independent events, the joint probability of all of the events occurring is equal to the product of the separate probabilities of the events.

In terms of three independent events A, B and C, the law can be expressed as:

$$P[A \cap B \cap C] = P[A] . P[B] . P[C]$$

(The symbol $\cap$ can be considered as representing the word 'and'.)

Other conventions used to write probability expressions are:

(i) $P[A \cap B \cap C]$ is often written as $P[ABC]$
(ii) $P[A] . P[B] . P[C]$ is usually written as $P[A]P[B]P[C]$.

Whenever there is no risk of ambiguity these later conventions are used in the text.

Example 1.8

Probability independence is illustrated by the tossing of dice:
If two dice are thrown what is the probability of two 3s?
Let $P[A]$ = probability of a 3 on the first die = $\frac{1}{6}$.
Let $P[B]$ = probability of a 3 on the second die = $\frac{1}{6}$.
Then probability of two 3s, $P[A \cap B] = \frac{1}{6} \times \frac{1}{6} = \frac{1}{36}$.

A further look at union probability

Considering the previous example. What is the union probability of either A or B (i.e. the probability of obtaining a 3 on either die or on both)?

If we use the summation law in the form stated above then:

$$P[A \cup B] = P[A] + P[B] = \tfrac{1}{6} + \tfrac{1}{6} = \tfrac{1}{3}$$

However, if we obtain this probability by enumeration we achieve a different value.

The set of events that cause a 3 to be scored on either die is a subset of the sample space shown in Fig. 1.1 and is:

$$(3, 1), \quad (3, 2) \quad (3, 3) \quad (3, 4) \quad (3, 5) \quad (3, 6)$$
$$(1, 3) \quad (2, 3) \quad (3, 3) \quad (4, 3) \quad (5, 3) \quad (6, 3)$$

A total of 12 events out of a total of 36 which gives a probability value of $\tfrac{12}{36}$ equalling the $\tfrac{1}{3}$ value obtained from the formula.

However, if these 12 events are examined we see that the event (3, 3), the probability of a 3 being scored on each die, has been included twice. There are only 11 events that involve at least one 3 being scored.

The true probability must be $\tfrac{11}{36}$.

The union probability formula is at fault because, in this problem, events A and B are independent and the joint probability that they may occur together, $P[A \cap B]$, has been included twice.

Things can be put right by simply subtracting $P[A \cap B]$ from the value so far obtained to give:

$$P[A \cup B] = P[A] + P[B] - P[A \cap B] = \tfrac{1}{3} - \tfrac{1}{36} = \tfrac{11}{36}$$

This is the general form of the summation law and applies to all events whether mutually exclusive or not. (If A and B are mutually exclusive then $P[A \cap B] = 0$.)

The complement of a probability

The complement of the probability of event A is given the symbol $P(\bar{A})$ and is the probability that event A will not occur.

$$\text{Now } 0 \leqslant P[A] \leqslant 1, \text{ hence } P[\bar{A}] = 1 - P[A]$$

If we have two sets A and B then, from the distributive law:

$$(A \cap \bar{B}) \cup (A \cap B) = A \cap (\bar{B} \cup B) = A \quad (\text{as } \bar{B} \cup B = 1.0)$$

which leads to another useful law of probability:

$$P[A \cap \bar{B}] = P[A] - P[A \cap B]$$

A convenient use for the complement of a probability is when it is required to estimate the probability of occurrence of a single event over a given number of trials.

Assume that, in one trial, the probability of occurrence of an event A is P[A]. Then, the probability of non-occurrence of A is $P[\bar{A}] = 1 - P[A]$ and the probability of occurrence of A in n trials $= 1 - (1 - P[A])^n$.

Example 1.9

Determine the probability of obtaining at least one '3' after six throws of a fair die.

Solution
Let P[A] be the probability of obtaining a '3' with one throw. Then:

$$P[A] = \tfrac{1}{6}$$

At first glance it would seem that if there is a one in six chance of obtaining the score '3' with one throw then the chance of obtaining at least one '3' after six throws of the die is $\tfrac{1}{6} + \tfrac{1}{6} + \tfrac{1}{6} + \tfrac{1}{6} + \tfrac{1}{6} + \tfrac{1}{6} = 1.0$, i.e. 100% or absolute certainty, a situation that does not fit reality.

The event of scoring a 3 can happen once in each throw and there are therefore six such possible events. These events are independent but not mutually exclusive and it is because of this that the simple addition procedure suggested above is not relevant.

For six throws of the die the probability of obtaining at least one score of 3 is given by the expression:

$$P = P[A] \cup P[A] \cup P[A] \cup P[A] \cup P[A] \cup P[A]$$

which, from the associative law, can be grouped in sets of twos:

$$P = P[A \cup A] \cup P[A \cup A] \cup P[A \cup A]$$
$$= P[B] \cup P[B] \cup P[B] \qquad \text{where } P[B] = P[A \cup A]$$
$$= P[A] + P[A]$$
$$- P[A]P[A]$$
$$= [\tfrac{2}{6} - \tfrac{1}{36}]$$
$$= \tfrac{11}{36}$$
$$= 0.3055$$

$$= P[C] \cup P[B] \qquad \text{where } P[C] = P[B \cup B]$$
$$= P[B] + P[B]$$
$$- P[B]P[B]$$
$$= [\tfrac{22}{36} - \tfrac{121}{1296}]$$
$$= 0.5177$$

$$= 0.5177 + 0.3055 - 0.5177 \times 0.3055 = 0.6651$$

Whilst 6 throws are manageable the procedure becomes more cumbersome as the number of throws increases.

With this type of problem it is often best to think in terms of the complement of the probability. Let $A_1, A_2, A_3, A_4, A_5, A_6$ be the events of scoring 3 in the first, second, third, fourth, fifth and sixth throws respectively. Then:

$$P[A_1] = P[A_2] = P[A_3] = P[A_4] = P[A_5] = P[A_6] = \tfrac{1}{6}$$

Now, P[at least one score of 3] + P(no scores of 3] = 1.0. Therefore:

$$P[\text{ at least one score of 3}] = 1 - P[\text{no scores of 3}]$$
$$= 1 - P[\bar{A}_1 . \bar{A}_2 . \bar{A}_3 . \bar{A}_4 . \bar{A}_5 . \bar{A}_6]$$
$$= 1 - \tfrac{5}{6} . \tfrac{5}{6} . \tfrac{5}{6} . \tfrac{5}{6} . \tfrac{5}{6} . \tfrac{5}{6}$$
$$= 0.6651$$

The law of large numbers

Predictions of the type illustrated in the above example can only be accurate if the die is perfectly balanced and has no bias towards any particular numbers. However, even with bias, provided an experiment can be repeated as often as required, it is possible to obtain a close approximation to the value of the probability of an event by the application of the law of large numbers.

This law states that if an experiment is performed often enough, then, in the long run, the observed relative frequency of an event is virtually equal to the probability of the event.

With the advent of micro computers the law of large numbers can now be applied to many probability problems. For instance it is a simple matter to write a computer program that simulates the resuts of six throws of a fair die.

With such a program the author was able to determine the probability of scoring 3, by expressing it as the relative frequency of the occurrence (i.e. the number of 3s scored divided by the number of times the die was thrown, six times). The results were as follows:

Number of throws	Number of 3s scored	Probability
After 1 set of six throws	1	1.0000
After 10 sets of six throws	7	0.7000
After 100 sets of six throws	60	0.6000
After 1000 sets of six throws	689	0.6890
After 10 000 sets of six throws	6 639	0.6639
After 100 000 sets of six throws	66 570	0.6657

Example 1.10

In a certain region the subgrade is predominantly a silty soil with the odd clay lens. The average size of these lenses is 750 m^2.

If on a site 100×150 m^2 a clay lens exists, what is the probability of encountering it in any one of eight boreholes, symmetrically placed over the site?

Solution

Let the probability of finding the lens in a borehole be P[F]. Then, P[F] can be estimated from the ratio of the two areas:

$$P[F] = \frac{750}{100 \times 150} = 0.05$$

Now, P[F̄] = 1 − P[F] which is the probability of not encountering the lens in one borehole. Hence if the lens does exist then the chance of encountering it in at least one of the eight boreholes is:

$$P[F] = 1 - (1 - 0.05)^8 = 0.337 \ldots \text{ say } 34\%$$

Reliability

The complement of a probability value is known as its reliability. This simply means that if a system has a probability of failure of 10% then it is said to be 90% reliable.

Example 1.11

In a reliability analysis for a proposed concrete retaining wall the following probabilities of failure were obtained.

Risk of bearing capacity failure $= P_b = 0.003$
Risk of overturning $= P_o = 0.001$
Risk of sliding failure $= P_s = 0.002$
Risk of structural concrete failure $= P_c = 0.0005$

Determine a value for P_f, the probability of failure of the wall.

Solution
Assuming that the various modes of failure are mutually exclusive it is obvious that the occurrence of any one of these failure events will result in the failure of the wall.

$$P_f = P_b \cup P_o \cup P_s \cup P_c$$
$$= P_b + P_o + P_s + P_c$$
$$- (P_bP_o + P_bP_s + P_bP_c + P_oP_s + P_oP_c + P_sP_c)$$
$$= 0.0065 - 0.000014 = 0.00649$$

An alternative solution may be worked.

$$P_f = P_b \cup P_o \cup P_s \cup P_c$$

and, therefore,

$$\bar{P}_f = \overline{P_b \cup P_o \cup P_s \cup P_o}$$

From De Morgan's Law, which states
$$\overline{P_b \cup P_o \cup P_s \cup P_c} = \bar{P}_b \cap \bar{P}_o \cap \bar{P}_s \cap \bar{P}_c:$$

$$\bar{P}_f = 0.997 \times 0.999 \times 0.998 \times 0.9995 = 0.99351$$

Now:

$$P_f = 1 - \bar{P}_f = 1 - 0.99352 = 0.00649$$

Conditional probability

When two events, A and B, are referred to as being independent it means that the happening of one of these events will have no effect on the probability of the happening of the other and $P[A \cap B] = P[A]P[B]$. However, there are many cases when the happening of one event can have a direct effect on the probability of the happening of the other. The events are dependent and $P[A \cap B]$ is no longer equal to $P[A]P[B]$.

A simple illustration is the drawing of an ace from a pack of 52 playing cards on the second draw.

Let the probability of drawing an ace on the first draw be P[A] and let the probability of drawing an ace on the second draw be P[B]. Then:

$$P[A] = \tfrac{4}{52} = \tfrac{1}{13}$$

and:

$$P[B] = \tfrac{4}{51} \quad \text{(if there was no ace on the first draw)}$$

$$P[B] = \tfrac{3}{51} \quad \text{(if there had been an ace on the first draw)}$$

In such a situation we are forced to use another symbol, P[B | A], in place of P[B] where P[B | A] represents the value of P[B] after, and knowing the result of, event A. P[B | A] is known as the conditional probability of B.

It is therefore best to express the probability of the happening of two events, A and B, either as:

$$P[AB] = P[A]P[B \mid A]$$

or as:

$$P[AB] = P[B]P[A \mid B] \quad \text{(since } P[BA] = P[AB])$$

Note the use of the convention of writing $P[A \cap B]$ as P[AB].

When we have a set of dependent events the probability that all these events will occur can be evaluated by the use of conditional probabilities. For example, for three events A, B and C:

$$P[ABC] = P[A]P[B \mid A]P[C \mid AB]$$

Which is the mathematical way of saying: 'The probability that events A, B and C will all happen is equal to the probability of event A multiplied by the probability of B, knowing the result of event A, multiplied by the probability of C knowing the result of events A and B'.

A further definition of independence

Obviously if the events A, B and C are independent then P[B | A] is equal to P[B] and P[C | AB] equals P[C] so that the formula becomes:

$$P[ABC] = P[A]P[B]P[C]$$

This leads to a further definition of independence. If A and B are two events such that P[B | A] equals P[B] then A and B are statistically independent.

Example 1.12

Determine the probability that, if two fair dice are thrown: (a) The first die will score a 3 and the second will score a 4 and show that these events are independent; (b) The total score will be not less than nine if the difference between the two die scores is not less than two and show that these results are dependent.

Solution

(a) Let events be: A, first die score is 3; B, second die score is 4. The sample space of the 36 possible scores is set out below.

$$
\begin{array}{cccccc}
6,1 & 6,2 & (6,3) & (6,4) & (6,5) & (6,6) \\
5,1 & 5,2 & 5,3 & (5,4) & (5,5) & (5,6) \\
4,1 & 4,2 & 4,3 & 4,4 & (4,5) & (4,6) \\
3,1 & 3,2 & 3,3 & 3,4 & 3,5 & (3,6) \\
2,1 & 2,2 & 2,3 & 2,4 & 2,5 & 2,6 \\
1,1 & 1,2 & 1,3 & 1,4 & 1,5 & 1,6 \\
\end{array}
$$

By enumeration it is seen that P[A] equals P[B] which equals $\frac{6}{36} = \frac{1}{6}$ and that P[A ∩ B] is $\frac{1}{36}$.

Note:

$$P[A \cap B] = P[A \mid B]P[B] = P[A]P[B]$$

(if A and B are independent). Also:

$$P[A]P[B] = \frac{1}{6} \cdot \frac{1}{6} = \frac{1}{36} \; (= P[A \mid B]P[B])$$

Hence:

$$P[A \mid B] = P[A]$$

i.e. events A and B are independent.

(b) Let the events be: A, the total score is not less than 9 (these events are in parentheses in the sample space shown above); B, the difference between die scores not less than 2 (these events are underlined in the above sample space).
    By enumeration:

$$P[A] = \frac{10}{36}$$

$$P[B] = \frac{20}{36}$$

P[A ∩ B] is the probability that both events A and B will occur, and can be found by enumeration to equal $\frac{4}{36}$. Now:

$$P[A \cap B] = P[A \mid B]P[B]$$

i.e.

$$\frac{4}{36} = P[A \mid B] \cdot \frac{20}{36}$$

i.e.

$$P[A \mid B] = \tfrac{2}{5} = P[A]$$

Events A and B are dependent.

We can check the value obtained for $P[A \mid B]$ by substitution in the union probability formula:

$$P[A \cup B] = P[A] + P[B] - P[A \mid B]P[B]$$

$$= \tfrac{10}{36} + \tfrac{20}{36} - \tfrac{2}{5} \times \tfrac{10}{36} = \tfrac{4}{36}$$

Example 1.13

For two events, A and B, $P[A]$ is $\tfrac{1}{2}$; $P[B]$ is $\tfrac{1}{4}$ and $P[A \cap B]$ is $\tfrac{1}{8}$. Determine the values of:

$$P[A \mid B]; \quad P[A \cup B]; \quad P[A \cap \bar{B}]; \quad P[A \mid \bar{B}]; \quad \text{and} \quad P[\bar{A} \mid \bar{B}]$$

Solution

$$P[A \mid B] = \frac{P[A \cap B]}{P[B]} = \tfrac{1}{8} \cdot \tfrac{4}{1} = \tfrac{1}{2}$$

$$P[A \cup B] = P[A] + P[B] - P[A \cap B] = \tfrac{1}{2} + \tfrac{1}{4} - \tfrac{1}{8} = \tfrac{5}{8}$$

$$P[A \cap \bar{B}] = P[A] - P[A \cap B] = \tfrac{1}{2} - \tfrac{1}{8} = \tfrac{3}{8}$$

$$P[A \mid \bar{B}] = \frac{P[A \cap \bar{B}]}{P[\bar{B}]} = \tfrac{3}{8} \cdot \tfrac{4}{3} = \tfrac{1}{2}$$

$$P[\bar{A} \mid \bar{B}] = \frac{P[\bar{A} \cap \bar{B}]}{P[\bar{B}]}$$

$$= \frac{P[\overline{A \cup B}]}{P[\bar{B}]} \quad \text{(De Morgan's law } \bar{A} \cap \bar{B} = \overline{A \cup B})$$

$$= \frac{1 - P[A \cup B]}{P[\bar{B}]} = \tfrac{3}{8} \cdot \tfrac{4}{3} = \tfrac{1}{2}$$

The theorem of total probability

With this theorem it is possible to use conditional probability values without knowing whether the events that affect these values have occurred or not.

Consider a set of events $B_1, B_2, B_3, \ldots, B_n$ which are both mutually exclusive and also collectively exhaustive (i.e. one of the events will occur). Then P[A], the probability of another event, A, can be expressed as:

$$P[A] = \sum_{i=1}^{n} P[A \mid B_i]P[B_i]$$

PROOF

The events B_i are collectively exhaustive, i.e.

$$B_1, B_2, B_3, \ldots, B_n = S$$

Now:

$$P[A] = P[A]P[S] \quad (\text{as } P[S] = \text{Total probability} = 1)$$
$$= P[A]P[B_1 \cup B_2 \cup B_3 \cup \ldots, B_n)$$
$$= P[AB_1] + P[AB_2] + P[AB_3] + \ldots, P[AB_n]$$

Figure 1.7 shows the Venn diagram representing the mutually exclusive and collectively exhaustive events $B_1, B_2, B_3, \ldots, B_n$. It can be seen that event A intersects these events so that the events $AB_1, AB_2, AB_3, \ldots, AB_n$ are also mutually exclusive. Hence:

$$P[A] = P[AB_1] + P[AB_2] + P[AB_3] + \ldots, P[AB_n]$$

proving that:

$$P[A] = \sum_{i=1}^{n} P[AB_i] = \sum_{i=1}^{n} P[A \mid B_i]P[B_i]$$

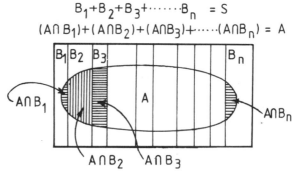

Fig. 1.7   Theory of total probability

The use of the total probability theorem can be illustrated by the simple example of selecting a red ball from a collection of red and white balls contained in four boxes, 1, 2, 3 and 4. Let us assume that:

Box 1 contains 2 red and 8 white balls
Box 2 contains 5 red and 5 white balls
Box 3 contains 6 red and 4 white balls
Box 4 contains 9 red and 1 white balls

We can easily work out P[A], the probability of selecting a red ball, provided that we know which box it has been taken from.

Let the events of selecting boxes 1, 2, 3 and 4 be B_1, B_2, B_3 and B_4 respectively. Then:

$$P[A \mid B_1] = 0.2; \quad P[A \mid B_2] = 0.5; \quad P[A \mid B_3] = 0.6; \quad P[A \mid B_4] = 0.9$$

However, if we do not know which box will be selected, there is further uncertainty and this can be allowed for by the total probability theorem. Assuming that the boxes have an equal chance of being selected:

$$P[B_1] = P[B_2] = P[B_3] = P[B_4] = 0.25$$

$$P[A] = \sum_{i=1}^{4} P[A \mid B_i]P[B_i]$$

$$= 0.2 \times 0.25 + 0.5 \times 0.25 + 0.6 \times 0.25 + 0.9 \times 0.25 = 0.55$$

Example 1.14

A building contractor requires a roll of roofing felt. There are three suppliers in the area and the probabilities (based on his previous experiences and the location of the suppliers) that the contractor will instruct his vanman to visit a particular supplier are:

A: the vanman goes to supplier A, $P[A] = 0.6$
B: the vanman goes to supplier B, $P[B] = 0.2$
C: the vanman goes to supplier C, $P[C] = 0.2$

Each supplier stocks roofing felt produced by two manufacturers, X and Y. Both types of roofing felt sell at the same price and both satisfy the current building regulations. The stock situation at each of the suppliers is:

Supplier	No. of 'X' rolls	No. of 'Y' rolls
A	10	30
B	30	20
C	30	10

The vanman will be told by his employer which supplier to visit.
Which roll type is the vanman most likely to return with?

Solution

Let P[X] be the probability that the vanman will return with roll type X. Then, considering the stock position of each supplier:

$$P[X \mid A] = \tfrac{10}{40} = 0.25$$

$$P[X \mid B] = \tfrac{30}{50} = 0.6$$

$$P[X \mid C] = \tfrac{30}{40} = 0.75$$

From the law of total probability:

$$P[X] = P[X \mid A]P[A] + P[X \mid B]P[B] + P[X \mid C]P[C]$$

$$= 0.25 \times 0.6 + 0.6 \times 0.2 + 0.75 \times 0.2$$

$$= 0.42$$

If P[Y] is the probability of obtaining type Y, it can be shown in a similar manner that P[Y] equals 0.58. Hence it is more likely that the vanman will return with a roll of type Y.

The tree diagram

The tree diagram, so named because of its appearance, can often be useful in decision making. Each branch illustrates the path that will be taken whenever a particular decision is made.

The tree diagram for example 1.14 is shown in Fig. 1.8 and shows all the possible combinations of events that could be involved in finishing up with an X or a Y type of roofing felt.

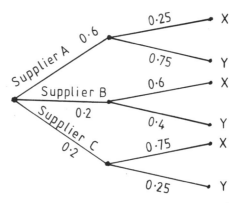

Fig. 1.8   Example 1.14

Tree diagrams are based on the theorem of total probability and can often be of assistance in a complex decision-making problem by representing it in a graphical form.

Example 1.15

If, in example 1.14, the vanman returned with an X type of roofing felt, determine the probability that he obtained it from supplier **B**.

Solution

The problem is simply to determine the value of $P[B \mid X]$.

$$P[B \mid X] = \frac{P[BX]}{P[X]}$$

Now:

$$P[BX] = P[XB] = P[X \mid B]P[B]$$

hence:

$$P[B \mid X] = \frac{P[X \mid B]P[B]}{P[X]}$$

$$= \frac{0.6 \times 0.2}{0.42} = 0.286$$

This is an application of Bayes' theorem, which gives a relationship between prior and posterior probabilities.

Bayes' theorem

We have seen that:

$$P[B_i \mid A] = \frac{P[B_iA]}{P[A]} = \frac{P[AB_i]}{P[A]}$$

Now:

$$P[A \mid B_i] = \frac{P[AB_i]}{P[B_i]}$$

hence:

$$P[B_i \mid A] = \frac{P[A \mid B_i]P[B_i]}{P[A]}$$

and, from the theorem of total probability:

$$P[A] = \sum_{i=1}^{n} P[A \mid B_i]P[B_i]$$

Therefore:

$$P[B_i \mid A] = \frac{P[A \mid B_i]P[B_i]}{\sum\limits_{i=1}^{n} P[A \mid B_i]P[B_i]}$$

The above expression is known as Bayes' theorem (or rule) and can be expressed in words as:

'If A is an event that could be caused by any one of n different events, B_i, all of which are both mutually exclusive and collectively exhaustive $\left(\sum\limits_{i=1}^{n} B_i = S\right)$, then Bayes' theorem gives the relationship between the probability of event B_i happening (given the result of the happening of event A) and the probability of A happening (given the result of the happening of event B_i).'

$P[B_i]$ = prior probability of event B_i (with no knowledge of A)

$P[B_i \mid A]$ = posterior probability of event B_i (after noting A)

$P[A \mid B_i]$ = likelihood of event A (after noting B_i)

Example 1.16

Solve example 1.15 using Bayes' theorem.

Solution

Let events A, B and C of example 1.14 be B_1, B_2 and B_3, so that $P[B \mid X]$ of example 1.15 is expressed as $P[B_2 \mid X]$. By Bayes' theorem:

$$P[B_2 \mid X] = \frac{P[X \mid B_2]P[B_2]}{\sum\limits_{i=1}^{3} P[X \mid B_i]P[B_i]}$$

$$= \frac{0.6 \times 0.2}{0.25 \times 0.6 + 0.6 \times 0.2 + 0.75 \times 0.2}$$

$$= 0.286$$

Note: Bayes' theorem is extremely useful as a technique to continually process information.

In any probability analysis the designer must make assumptions in order to have a set of prior probabilities. Bayes' theorem can be used to continually adjust these assumed probability values to conform with new information as it becomes available and is then usually written in the form:

$$P[\text{state} \mid \text{sample}] = \frac{P[\text{sample} \mid \text{state}]P[\text{state}]}{\sum\limits_{\text{all states}} P[\text{sample} \mid \text{state}]P[\text{state}]}$$

Example 1.17

The strength of an existing earth embankment is to be assessed.

The soil in the embankment is largely cohesive and, having studied records of its past performance and having visited the site, to check its appearance and general state of repair, the soils engineer has concluded that the soil is mainly of a firm consistency. His definitions of consistency were based on the unconfined compressive strength of the soil, c_u, as follows:

Soft consistency	$c_u < 24$ kN/m^2
Firm	$c_u = 24$ to 48 kN/m^2
Stiff	$c_u > 48$ kN/m^2

If we let B_i equal the average consistency of the soil in the embankment then B_i can be soft, firm or stiff. The engineer assigned prior probabilities for B_i as:

Average consistency (B_i)	Prior probability $P[B_i]$
Soft	0.3
Firm	0.5
Stiff	0.2

If the state of the soil in the embankment is known then it is a simple matter to predict a value for the unconfined strength, A, that one would obtain from a test on a sample. However one test can never be conclusive and the soils engineer, with the help of control tests and his experience with similar problems, evolved a set of conditional probabilities, $P[A \mid B_i]$, that are set out below.

A, the c_u value obtained in a test on a sample (kN/m^2)	B_i, the average state, i.e. consistency of the soil		
	Soft	Firm	Stiff
<24 (indicates soft, A_s)	0.7	0.3	0
24 to 48 (indicates firm, A_f)	0.3	0.6	0.2
>48 (indicates stiff, A_{st})	0	0.1	0.8

He then slightly adjusted the $P[A \mid B_i]$ values to use 0.01 instead of 0 and hence avoid any multiplications by zero. This is necessary in order to avoid the elimination of a probability that may increase as more data becomes available.

Sample (A_i)	State ($P[A_i \mid B_i]$)		
	Soft	Firm	Stiff
A_s	0.7	0.3	0.01
A_f	0.29	0.6	0.19
A_{st}	0.01	0.1	0.8

If the reader is in some doubt as to what, exactly, $P[A_i \mid B_i]$ means perhaps the following explanation may be of asistance.

If the state is soft, i.e. B_i equals 'soft', then the probability of a test on a sample indicating this situation (i.e. recording an unconfined compressive strength of less than 24 less than kN/m^2) is 0.7. In symbols:

$$P[A_s \mid B_s] = 0.7$$

$$P[A_s \mid B_f] = 0.3$$

$$P[A_s \mid B_{st}] = 0.01$$

where the subscripts s, f and st stand for soft, firm and stiff respectively.

Obviously, for a particular state, a test result must reflect some consistency, namely soft, firm or stiff, which explains why the $P[A \mid B_i]$ values in the table summate vertically to 1.0.

Assume that, in this case, four samples were collected from different locations in the embankment and that by the unconfined compression test two of the samples indicated a stiff consistency, A_{st}, one a firm consistency, A_f, and one a soft consistency, A_s. Determine the main consistency state of the embankment.

Solution
The prior probabilities, assumed by the engineer, for the state of the embankment are:

$$P[B_s] = 0.3; \quad P[B_f] = 0.5; \quad P[B_{st}] = 0.2$$

Consider test 1, result A_{st}.

Having a test result, A_{st}, we can work out the posterior probabilities as to the state of the embankment:

$P[B_s \mid A_{st}]$ = the probability that the consistency is soft (knowing that the test result indicates that it is stiff)

$P[B_f \mid A_{st}]$ = the probability that the consistency is firm

$P[B_{st} \mid A_{st}]$ = the probability that the consistency is stiff

From Bayes' theorem:

$$P[B_s \mid A_{st}] = \frac{P[A_{st} \mid B_s]P[B_s]}{P[A_{st} \mid B_s]P[B_s] + P[A_{st} \mid B_f]P[B_f] + P[A_{st} \mid B_{st}]P[B_{st}]}$$

$$= \frac{0.01 \times 0.3}{0.01 \times 0.3 + 0.1 \times 0.5 + 0.8 \times 0.2}$$

$$= \frac{0.003}{0.213} = 0.0141$$

$$P[B_f \mid A_{st}] = \frac{0.1 \times 0.5}{0.213} = 0.2347$$

$$P[B_{st} \mid A_{st}] = \frac{0.8 \times 0.2}{0.213} = 0.7512$$

Hence, prior to considering the results of the second test, we have upgraded the $P[B_i]$ values to:

$$P[B_s] = 0.0141; \quad P[B_f] = 0.2347; \quad P[B_{st}] = 0.7512$$

The process continues in an identical manner:

Test 2, result A_{st}

$$P[B_s \mid A_{st}] = \frac{0.01 \times 0.0141}{0.01 \times 0.0141 + 0.1 \times 0.2347 + 0.8 \times 0.7512}$$

$$= \frac{0.0001}{0.6246} = 0.0002$$

$$P[B_f \mid A_{st}] = \frac{0.1 \times 0.2347}{0.6246} = 0.0376$$

$$P[B_{st} \mid A_{st}] = \frac{0.8 \times 0.7512}{0.7512} = 0.9622$$

Test 3, result A_f

$$P[B_s \mid A_f] = \frac{0.29 \times 0.0002}{0.29 \times 0.0002 + 0.6 \times 0.0376 + 0.19 \times 0.9622}$$

$$= \frac{0.00006}{0.2054} = 0.0002$$

$$P[B_f \mid A_f] = \frac{0.6 \times 0.0376}{0.2054} = 0.1097$$

$$P[B_{st} \mid A_f] = \frac{0.19 \times 0.9622}{0.2054} = 0.8900$$

Test 4, result A_s

$$P[B_s \mid A_s] = \frac{0.7 \times 0.0002}{0.7 \times 0.0002 + 0.3 \times 0.1097 + 0.01 \times 0.8900}$$

$$= \frac{0.00014}{0.0420} = 0.0033$$

$$P[B_f \mid A_s] = \frac{0.3 \times 0.1097}{0.0420} = 0.7845$$

$$P[B_{st} \mid A_s] = \frac{0.01 \times 0.8900}{0.0420} = 0.2122$$

Probabilities of state of embankment are:

Soft consistency	$P =$	0.3%
Firm consistency	$P =$	78.5%
Stiff consistency	$P =$	21.2%

The reader might like to check that the final probabilities are unaffected by the order in which the test results are considered.

Alternative solution
The foregoing procedure was listed in full as a demonstration but it is not necessary to consider each test result separately.

As the test results are independent, the values of the conditional probabilities of the results, $P[A \mid B_i]$, are equal to the product of the four conditional probabilities:

$$P[A \mid B_s] = P[A_{st} \mid B_s]P[A_{st} \mid B_s]P[A_f \mid B_s]P[A_s \mid B_s]$$

$$= 0.01 \times 0.01 \times 0.29 \times 0.7 = 0.0000203$$

$$P[A \mid B_f] = P[A_{st} \mid B_f]P[A_{st} \mid B_f]P[A_f \mid B_f]P[A_s \mid B_f]$$

$$= 0.1 \times 0.1 \times 0.6 \times 0.3 = 0.0018$$

$$P[A \mid B_{st}] = P[A_{st} \mid B_{st}]P[A_{st} \mid B_{st}]P[A_f \mid B_{st}]P[A_s \mid B_{st}]$$

$$= 0.8 \times 0.8 \times 0.19 \times 0.01 = 0.001216$$

Now, from Bayes' theorem:

$$P[B_s \mid A] = \frac{P[A \mid B_s]P[B_s]}{P[A \mid B_s]P[B_s] + P[A \mid B_f]P[B_f] + P[A \mid B_{st}]P[B_{st}]}$$

The prior probabilities for B_s, B_f and B_{st} were 0.3, 0.5 and 0.2. Hence:

$$P[B_s \mid A] = \frac{0.0000203 \times 0.3}{0.0000203 \times 0.3 + 0.0018 \times 0.5 + 0.001216 \times 0.2}$$

$$= 0.0053$$

and:

$$P[B_s \mid A] = \frac{0.0018 \times 0.5}{0.001149} = 0.7831$$

$$P[B_{st} \mid A] = \frac{0.00122 \times 0.2}{0.001149} = 0.2116$$

giving:

$$P[B_s] = 0.5\%; \qquad P[B_f] = 78.3\%; \qquad P[B_{st}] = 21.2\%$$

Note: Example 1.17 indicates the possible dangers of adopting a deterministic approach to solving a civil engineering problem.

The soils engineer is experienced, knows the site, the testing techniques and the personnel involved. His judgement has been included in a probability analysis which indicates strongly that the soil in the embankment is almost entirely of a firm consistency.

If the test results are considered then there is every likelihood that the conclusion will be that the embankment consists predominantly of stiff soil.

Example 1.18

In the clay lens problem of example 1.10 the engineer assesses that there is a 50% chance of a clay lens being within the site area. He also estimates that, if there is a lens, the chance of encountering it in a borehole is 0.05 (see method in example 1.10).

Determine the change in this probability if, after eight boreholes, no clay has been encountered.

Solution
There are two states: 1, lenses present; 2, no lenses are present. And, there are two test results (samples): 1, no find; 2, a find.

State – lenses present
The probability of a find in one borehole, P[find], is 0.05. Hence:

$$P[\text{no find}] = 1 - P[\text{find}] = 1 - 0.05 = 0.95$$

Hence the probability of no find in eight boreholes is 0.95^8.

State – no lenses present
Obviously P[find] is 0 and P[no find] is 1. Using Bayes' theorem:

P[lenses | no find] =

$$\frac{P[\text{no find} \mid \text{lenses}] \times P[\text{lenses}]}{P[\text{no find} \mid \text{lenses}] \times P[\text{lenses}] + P[\text{no find} \mid \text{no lenses}] \times P[\text{no lenses}]}$$

$$= \frac{0.95^8 \times 0.5}{0.95^8 \times 0.5 + 1 \times 0.5} = 0.399$$

The probability of there being clay lenses within the site area has reduced from 50 to 40%.

Note: The same result is obtained if the calculation is carried out for the five separate events of not finding clay in each borehole in a similar manner to the way example 1.17 was first solved. However, apart from the extra work involved, it is necessary to work to several places of decimal to avoid significant rounding-off errors.

Exercises

SETS

1.1 A = (1, 3, 5, 7, 9, 11, 13) and B = (5, 7, 9)
Evaluate $A \cap B$; $A \cup B$ and $\bar{B}_A$.

Answer: (5, 7, 9); (1, 3, 5, 7, 9, 11, 13); (1, 3, 11, 13)

1.2 A = (2, 4, 6, 8, 10, 12) and B = (2, 3, 4, 5, 7, 8)
Evaluate $(A - B)$; $(A \cap B)$; $(A \cup B)$.

Answer: (6, 10, 12); (2, 4, 8); (2, 3, 4, 5, 6, 7, 8, 10, 12)

1.3 Show, by means of a Venn diagram, that if A, B and C are three sets, that:
  (i) $A \cup (B \cup C) = A \cup B \cup C$
  (ii) $A - B = A - \bar{B}$
  (iii) $\bar{A} \cap \bar{B} = \overline{A \cup B}$

1.4 Solve 1.3 numerically for:
A = (6, 9, 12, 15, 18, 21)
B = (5, 7, 8, 10, 13)
C = (2, 4, 6, 8, 10)

Hint for parts (ii) and (iii): sample space, $S = A \cup B \cup C$ and $\bar{B} = S - B$.

PROBABILITY

1.5 A card is drawn randomly from a pack of 52 playing cards. Determine the probability that:
  (i) The card chosen is black.
  (ii) The card chosen is a black, Ace, King, Queen or Jack.

Answer: (i) 0.5; (ii) 0.154

1.6 On a particular building site the lorries delivering materials are fuelled either by petrol or diesel.
    60% of the lorries deliver sand and 55% of the lorries use petrol.
    Determine the probability that a lorry chosen at random will not be delivering sand and will be diesel powered.

Answer: 0.18

1.7   (i) Determine the probability of scoring six with one throw of two fair dice.

(ii) Determine the probability of scoring at least one six with three throws of the dice

Answer: (i) 0.139; (ii) 0.362

Hint: use Fig. 1.1 for (i) and use the complement probability for (ii).

1.8   In previous years piling contractor A has successfully tendered for 8 out of 20 piling contracts whereas piling contractor B obtained 6 of the same 20 contracts.

The contractors have both been asked to tender for three piling contracts. On the basis of previous form determine the probability that:

(i) Contractor A will obtain at least one contract.

(ii) Contractor B will obtain at least one contract.

(iii) Either contractor A or contractor B will obtain at least one contract.

(iv) Contractor A will obtain all three contracts.

Hint: in (i) probability $= P[A \cup A \cup A]$.

Answer: (i) 0.784; (ii) 0.657; (iii) 0.973; (iv) 0.064

DEPENDENCY

1.9   A and B are independent events such that $P[A \cup B] = 0.58$ and $P[A \cap B] = 0.12$.

Determine two possible values for $P[A]$.

Answer: 0.4 or 0.3

1.10   A and B are two events such that $P[A] = \frac{1}{3}$; $P[B] = \frac{2}{3}$ and $P[A \cup B] = \frac{1}{2}$.

(i) Show that A and B are dependent events.

(ii) Determine the value of $P[A \mid B]$.

Answer: (i) A and B are dependent as $P[A \mid B] \neq P[A]$; (ii) $\frac{3}{4}$

TOTAL PROBABILITY

1.11   A foundation will be subjected to two dependent loadings, A and B, which can both vary in magnitude from 200 to 800 kN.

The design engineer has estimated values for $P[B_i]$, the probabilities of B achieving magnitudes of 200, 400, 600 and 800 kN and for $P[A_i \mid B_i]$, the conditional probabilities of A achieving the same magnitudes.

These values are set out in the table.

	$P[A_i \mid B_i]$					$P[B_i]$
	A_{200}	A_{400}	A_{600}	A_{800}	$\sum P[A_i \mid B_i]$	
B_{200}	0.1	0.5	0.3	0.1	1.0	0.3
B_{400}	0.3	0.5	0.15	0.05	1.0	0.4
B_{600}	0.6	0.25	0.1	0.05	1.0	0.2
B_{800}	0.7	0.15	0.1	0.05	1.0	0.1

$$\sum P[B_i] = 1.0$$

(i) Determine the probabilities that A will achieve 200, 400, 600 and 800 kN.

Hint: for the probability of A achieving 400, i.e. $P[A_{400}]$, working is as set out below.

$$P[A_{400}] = \sum_{i=200}^{800} P[A_{400} \mid B_i]P[B_i]$$

$$= 0.3 \times 0.3 + 0.5 \times 0.4 + 0.25 \times 0.2 + 0.15 \times 0.1$$

$$= 0.355$$

(ii) Determine the probabilities that the total load will be equal to: 400 kN; 600 kN; 800 kN; 1000 kN; 1200 kN; 1400 kN; 1600 kN.

Hint: for the 1000 kN prediction the working is as follows:
Let $(L = 1000)$ be the event that the total foundation load equals 1000 kN. Then:

$$(L = 1000) = (B_{200} \cap A_{800}) \cup (B_{400} \cap A_{600})$$

$$\cup (B_{600} \cap A_{400}) \cup (B_{800} \cap A_{200})$$

Hence:

$$P(L = 1000) = P[B_{200} \cap A_{800}] \cup P[B_{400} \cap A_{400}]$$
$$\cup P[B_{600} \cap A_{400}] \cup P[B_{800} \cap A_{200}]$$

$$= P[A_{800} \mid B_{200}]P[B_{200}]$$

$$+ P[A_{600} \mid B_{400}]P[B_{400}]$$

$$+ P[A_{400} \mid B_{600}]P[B_{600}]$$

$+ P[A_{200} \mid B_{800}]P[B_{800}]$

$$= 0.1 \times 0.3 + 0.15 \times 0.4 + 0.25 \times 0.2$$

$$+ 0.7 \times 0.1$$

$$= 0.21$$

Answer: (i) 0.34; 0.355; 0.24; 0.065. (ii) 400 kN – 0.03; 600 kN – 0.21; 800 kN – 0.47; 1000 kN – 0.21; 1200 kN – 0.055; 1400 kN – 0.02; 1600 kN – 0.005.

Note: Although it is always satisfactory to obtain the right answer to a question, readers are advised not to spend too much time on any exercise with which they find difficulty. An understanding of what probability is and how one can estimate probability values is really all that is required in order to work through the reliability analyses presented in later chapters.

Chapter Two
Random Variables

Random variables

In order to use probability theory it is necessary to express engineering uncertainties in terms of numerical values which can then be considered to vary in an uncertain, or random, way.

For example if, on a particular site, the unit weight of the soil varies between 18 and 20 kN/m³ it is not possible to fix an actual numerical value for this parameter.

The procedure adopted is to designate such fluctuating values as capital letters A, B, C, ... etc. signifying that the particular value of the parameter represented by the letter is not constant but varies randomly over a range of possible values, in our case from 18 to 20 kN/m³.

Lower case letters a, b, c, ... are generally used to denote the various values that the random variables A, B, C, ... can have.

Stochastic variables

A particular form of the random process is the stochastic process in which the frequency of occurrence of different values of a variable follows some form of statistical pattern. In such a case the variable, while still correctly described as random, is sometimes referred to as a stochastic variable.

The variables connected with civil engineering are almost always stochastic so that if the relevant variables for a particular civil engineering structure can be identified, and their probability laws obtained, then predictions as to the probable behaviour of the structure become possible. However, the probability law obeyed by a variable is often not known and it must therefore be estimated by some statistical method.

The frequenistic approach, described in chapter 1, in which the future behaviour of a variable is predicted from its past behaviour, is, generally, the method most suitable for civil engineering materials.

A series of tests, measuring the required parameter, is carried out on samples of the material and the results then analysed on the premise that the behaviour of the mass can be extrapolated from the behaviour of samples taken from the mass.

Just as a set of triaxial test results can be used to assess a value for the strength of a soil mass, so the answers given by a cross-section of housewives in a given area, when asked about a certain product, can be used to assess the average feeling of all housewives in that region towards the product. The sampling procedure used in both cases must of course be realistic. The soil samples collected for testing should be representative of the soil deposit just as the housewives interviewed should be representative of other housewives in the district.

In civil engineering the term, 'sample' means a single item, such as a soil sample or a single concrete test cube.

In statistics the term 'sample' means a set of results or values.

Generally, n samples of a civil engineering material will yield one statistical sample consisting of n values of the parameter measured with each sample of the material.

Discrete and continuous random variables

A random variable is classified by the form that its possible values will take.

If the values of the variable can only be from a distribution of finite values, e.g. the number of times an automatic valve operates in a twenty-four hour period, the variable is referred to as a discrete random variable.

If the values of the variable are continuous, e.g. 6.00001 is considered different from 6.00002, then the variable is referred to as a continuous random variable.

Most engineering situations involve continuous variables. However, in many cases, the process of rounding off to so many decimal places or the limitations of the measuring apparatus results in a set of measured values that appear to be from a discrete distribution.

Graphical representation of values of a variable

A particular measurement or value of a variable is best regarded as a random sample taken from the full range of possible values. For instance, measurements of unit weight obtained from a compacted earth fill are values of the random variable X where X is the random variable representing the unit weight of the fill.

In many cases a mass of numerical data is almost unintelligible. However, if the numbers are values of a stochastic variable then some form of graphical representation can often be useful. The most suitable form of graphical presentation depends on whether the variable is either discrete or continuous.

GRAPHICAL FORMS FOR DISCRETE RANDOM VARIABLES

Further examples of discrete random variables are:

(i)   The number of parking tickets issued daily by a traffic warden.
(ii) The number of aircraft waiting to take off at an airport.
(iii) The number of votes cast for a particular election candidate.

No doubt the reader can think of many others.

Cases such as (iii) and possibly (i) where the number of values involved may be large can often be treated as continuous random variables – there is little difference between 19 058 and 19 057 votes. However, the variable is still discrete in that it is not possible for a candidate to receive part of a vote, 19 058.327 for example.

Example (iii) also illustrates the one-off experiment or simple event. Whereas the number of parking tickets issued will vary from day to day, the number of votes the candidate obtains is for one event and, although the range of possible votes extends from zero to the number of voters partaking in the election, only one value is possible. In this case any meaningful forecast of the result can only be obtained from probability theory using prior probability values estimated from door-to-door canvassing, knowledge of the candidate's previous record, etc.

(A) THE LINE DIAGRAM

A line diagram (or bar chart) can be used to create a graphical representation by which the range and distribution of the measured values of a discrete random variable are easily seen.

The data is first placed into some order, usually of either increasing or decreasing values. It is then a fairly simple task to determine the number of times that a particular value (or group of values) occurs. Such a table is known as a frequency distribution table.

Example 2.1

The work of a lollipop lady in charge of a primary school crossing was monitored one morning. The group size, i.e. the number of children in each group that she supervised across the road, was noted as well as the number of times each group size occurred. The survey yielded the following frequency distribution table:

Group size	1	2	3	4	5	6	7	8	9	10	⩾11
Occurrence	2	1	3	4	3	5	4	3	2	1	0

Using a line diagram plot the group frequency distribution.

Solution
The line diagram is shown in Fig. 2.1.

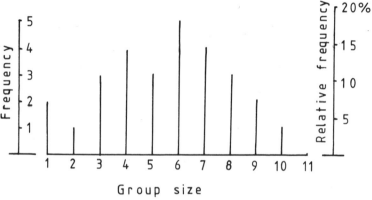

Fig. 2.1   Line diagram for example 2.1

(B) RELATIVE FREQUENCY DISTRIBUTION

It is sometimes more convenient to express the length of the lines of a line diagram as a proportion (or percentage) of the total number of values.

In example 2.1 the total number of groups was 28 so that, for the group size 6, the proportional number of groups is $\frac{5}{28}$ or 0.179 (or 17.9%).

When plotted in this form the diagram is the same as in Fig. 2.1 but the vertical scale is now such that the total length of the lines, if added together, equals 1.0. A diagram of this form is called the relative frequency distribution.

(C) THE CUMULATIVE FREQUENCY DIAGRAM

Sometimes referred to as an 'ogive', this is another widely-used method for the presentation of discrete data in graphical form. By means of this curve it is possible to tell at a glance the number of values that are equal to or are less than any particular value.

The procedure is to select particular values, ranging from the low to high end of the range, and to plot the number of values in the data that are equal to or are less than each selected value.

Associated with the cumulative frequency curve is the relative frequency cumulative curve, obtained by dividing each cumulative total by the total number of observations.

Example 2.2

Plot the cumulative frequency curve for example 2.1.

Solution

The cumulative frequency distribution table is set out below:

Group size	Group frequency	Cumulative	Relative cumulative frequency
1	2	2	$\frac{2}{28} = 0.071$
2	1	3	0.107
3	3	6	0.214
4	4	10	0.357
5	3	13	0.464
6	5	18	0.643
7	4	22	0.786
8	3	25	0.893
9	2	27	0.964
10	1	28	1.000
11	0	28	1.000

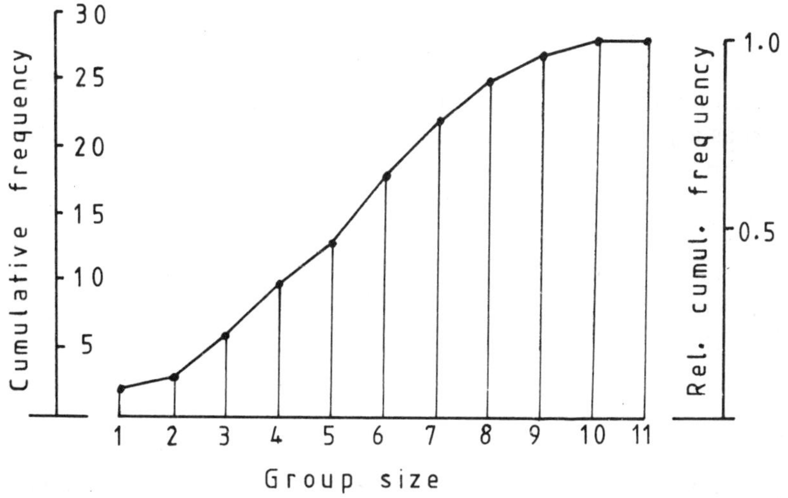

Fig. 2.2 Cumulative frequency curve for example 2.2

The cumulative frequency curve is shown in Fig. 2.2. The scale representing the relative cumulative frequency curve is also plotted on the figure.

GRAPHICAL FORMS FOR CONTINUOUS VARIABLES

When dealing with a discrete random variable, such as X where X equals the values of the integers from 1 to 10, it is possible to write down the ten values that X can have.

If, however, X is a continuous random variable it can have any value between 1 and 10 (where 'any' implies that there is no limit on the number of figures behind the decimal point). It is impossible to write down the total possible values of X.

Obviously the methods just described for discrete variables cannot be applied directly to continuous variables.

The histogram is a popular diagram used in statistics for the presentation of data obtained for a continuous variable. It is a diagram in which the range of values measured is divided into a number of cells (usually equal) which each represent a small portion of the range. When a set of apparently scattered measurements is presented in this form, an indication of the distribution of these measurements often emerges.

Example 2.3

A series of unconfined compression tests was carried out on undisturbed samples taken from a soil stratum. The values of the measured strengths are given in Table 2.1.

Table 2.1   Example 2.3 – measurements of c_u (kN/m^2)

37.6	35.9	34.9	34.0	32.8	30.4	28.5	28.9	31.4
33.3	27.5	29.6	33.9	38.1	34.5	34.4	32.7	36.8
39.9	31.7	34.1	31.1	27.4	33.4	29.8	29.2	25.3
31.4	27.1	31.8	29.1	31.3	35.8	36.2	27.4	29.3
24.8	26.0	30.5	38.4	37.3	39.7	32.5	34.5	27.5

Plot the measured values in the form of a histogram.

Solution
The minimum value measured was 24.8 kN/m^2; the maximum value measured was 39.9 kN/m^2.

A sensible range of values for the construction of the histogram would appear to be from 24 to 40 kN/m^2. If we select five cells then their widths will equal 3.2 kN/m^2 and will give the histogram shown in Fig. 2.3a. If, however, we halve the cell width, i.e. use 1.6 kN/m^2, then there will be ten cells, and the histogram will be as shown in Fig. 2.3b.

If there is difficulty in choosing a suitable cell number Benjamin and Cornell (1970) quote an empirical rule that was suggested some years ago:

$$k = 1 + 3.3 \log_{10} n$$

where:

n = number of data values
k = number of cells between maximum and minimum values

With this rule $k = 6.5$ which, if made up to 7, gives a cell width of $(40 - 24)/7 = 2.28$ kN/m^2. Rounding off this cell width to 2.3 kN/m^2 and using seven cells gives a range between minimum and maximum values of 24 to 40.1 kN/m^2 and the histogram with this arrangement is shown in Fig. 2.3c.

In order to draw these histograms the number of values that occur in each cell, i.e. the cell frequency, must be obtained. A convenient way to carry out this task is demonstrated in Table 2.2 which lists the data values of the example and is the procedure necessary for the seven-celled histogram. A vertical stroke is made to record each of the first four values found and a horizontal line is drawn through them in order to record a fifth value.

A table such as Table 2.2 is known as a frequency distribution table.

Various assumptions can be made whenever a value falls exactly on a boundary, values 38.4 and 30.4 in the example. In this case values on a boundary were assigned to the lower cell.

Table 2.2 Example 2.3

Range	Frequency
24 –26.3	/// = 3
26.31–28.6	~~////~~ / = 6
28.61–30.9	~~////~~ /// = 8
30.91–33.2	~~////~~ //// = 9
33.21–35.5	~~////~~ //// = 9
35.51–37.8	~~////~~ / = 6
37.81–40.1	//// = 4

From a study of Fig. 2.3 it is seen that, unlike the raw data of Table 2.1, the histogram gives an impression of the range of the data and an idea as to how they are distributed within that range.

However, Fig. 2.3 also illustrates just how histograms created from the same data vary in shape with each different choice of cell width.

The area of a histogram is of little importance. What is important is that, provided the cells are all of the same width, the total area under the resulting histogram divided by the cell width equals the total number of observations.

It is up to the reader to choose a cell width that creates the histogram best suited to his needs.

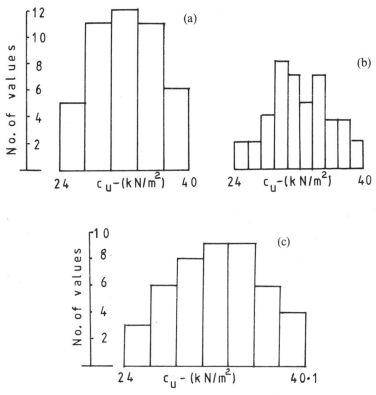

Fig. 2.3   Histograms for example 2.3

THE FREQUENCY POLYGON

Another method of presenting the data obtained for a continuous random variable is the frequency polygon which is the line joining the midpoints of each column of the histogram. The diagram is usually extended by one cell width on each side and the plotted line is drawn down to the midpoints of these two extra cells.

The frequency polygons corresponding to the histograms shown in Figs 2.3a, b and c are shown in Figs 2.4a, b and c respectively.

It should be noted, however, that a frequency polygon can be drawn directly from a frequency distribution table, such as Table 2.2, without any reference to a histogram.

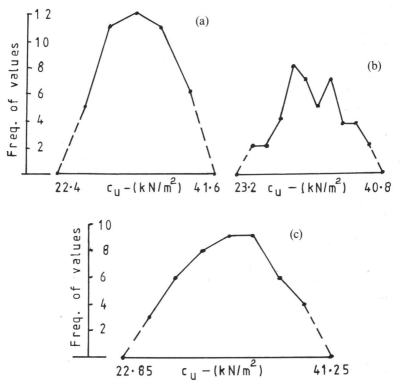

Fig. 2.4 Frequency polygons for example 2.3

THE CUMULATIVE FREQUENCY POLYGON

This diagram is similar to that for a discrete random variable. It is used to observe the total number of values within the distribution that are equal to or less than the maximum value of a particular cell.

The cumulative frequency distribution, in which the total number of measured values is represented by 1.0 or 100%, is obtained by considering each cell in turn, in ascending order, and dividing the summation of the number of values recorded for the cell, together with those below it, by the total number of values.

Example 2.4

Using seven cells and the data from example 2.3 plot the cumulative frequency distribution.

Solution
The cumulative frequency plot can be obtained from the information contained in Table 2.2 which leads to Table 2.3

Table 2.3   Cumulative frequency distribution for example 2.4

Cell range (kN/m²)	Cumulative frequency (C)	Relative cumulative frequency ($C/N \times 100\%$)
24 −26.3	3	6.7
26.31–28.6	9	20.0
28.61–30.9	17	37.8
30.91–33.2	26	57.8
33.21–35.5	35	77.8
35.51–37.8	41	91.1
37.81–40.1	45	100.0

The cumulative frequency distribution is shown in Fig. 2.5.

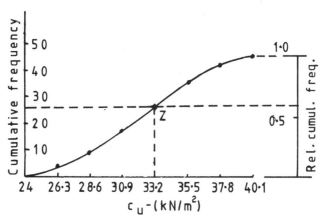

Fig. 2.5   Plot for example 2.4

Assume that we wish to know how many of the measurements are less than or equal to 33.2 kN/m².

In Fig. 2.5 draw a vertical line up from 33.2 kN/m² to cut the plotted line at point Z. From Z draw a horizontal line to cut the vertical scales. The intersections with these scales indicate that the number of values less than or equal to 33.2 kN/m² is 26, or 57.8% of the values are less than 33.2 kN/m².

Probability distributions

One of the major properties of a stochastic variable is that it is possible to estimate the likelihood, or the probability, that it will achieve some particular value.

A convenient way to obtain this probability is to study the mathematical function known as the probability distribution of the variable. Such a function is either discrete or continuous, depending on the character of the random variable involved.

The probability distribution function for a discrete variable is referred to as a probability mass function (pmf) and, for a continuous variable, as a probability density function (pdf).

THE PROBABILITY MASS FUNCTION (pmf)

If X is a discrete random variable capable of having n values, x_i, where i ranges from 1 to n, then $p_X(x_i)$ is the probability that X will equal a particular value x_i.

The probability mass function of X is therefore:

$$p_X(x_i) = P[X = x_i]$$

When there is no risk of ambiguity the subscript 'i', although still inferred, is dropped and the pmf of X is written as $p_X(x)$ where x represents any of the values possible for X. This convention is used throughout the book and the pmf of X is therefore written as:

$$p_X(x) = P[X = x]$$

Obviously the pmf must satisfy the laws of probability established in chapter 1, viz:

(i) $0 \leqslant p_X(x) \leqslant 1$ for all possible values of x
(ii) $\sum p_X(x) = 1$ with the summation taken for all possible values of x

Example 2.5

Determine the probability mass function for the scores that can be obtained by the throwing of two fair dice.

Solution
The problem is best tackled by enumeration and a tabulation of how all possible dice scores are formed is given in Table 2.3.

Table 2.3

Score	Combination					
2	1, 1					
3	1, 2	2, 1				
4	1, 3	2, 2	3, 1			
5	1, 4	2, 3	3, 2	4, 1		
6	1, 5	2, 4	3, 3	4, 2	5, 1	
7	1, 6	2, 5	3, 4	4, 3	5, 2	6, 1
8	2, 6	3, 5	4, 4	5, 3	6, 2	
9	3, 6	4, 5	5, 4	6, 3		
10	4, 6	5, 5	6, 4			
11	5, 6	6, 5				
12	6, 6					

Let X be the dice score. Then, from Table 2.3, we see that:

$$P[X = 2] = \tfrac{1}{36}; \qquad P[X = 3] = \tfrac{2}{36}, \quad P[X = 4] = \tfrac{3}{36}; \qquad \ldots \text{etc.}$$

Hence the probability mass function is:

$$p_X(x) = \begin{cases} 0.028 \\ 0.056 \\ 0.083 \\ 0.111 \\ 0.139 \\ 0.167 \end{cases} \text{for} \begin{cases} X = 2 \text{ or } X = 12 \\ X = 3 \text{ or } X = 11 \\ X = 4 \text{ or } X = 10 \\ X = 5 \text{ or } X = 9 \\ X = 6 \text{ or } X = 8 \\ X = 7 \end{cases}$$

The answer is to bet on the 7!

The pmf can be plotted as a line diagram and is shown in Fig. 2.6a.

THE CUMULATIVE DISTRIBUTION FUNCTION (cdf) FOR A DISCRETE VARIABLE

The cumulative distribution function is an alternative form of the probability distribution and it is often convenient to use it.

The cdf of X is written as $F_X(x)$ and is the probability that the actual value of the random variable, X, is equal to or is less than x, where x is some specified value of X within the range of possible values, i.e.

$$F_X(x) = P[X \leqslant x]$$

For a discrete random variable, X, the cdf is simply the sum of the probabilities of all the possible values of X that are equal to or are less than the argument, i.e. equal to or less than x_k, the value of X being considered.

$$F_X(x_k) = \sum_{i=1}^{i=k} p_X(x_i)$$

The cdf, $F_X(x)$, for the dice score of example 2.5 is shown in Fig. 2.6b. It should be noted that the value of $F_X(x)$ for a particular integer corresponds to the higher step. For example, for a score of 7, the cumulative frequency distribution is equal to 0.583 and not 0.417.

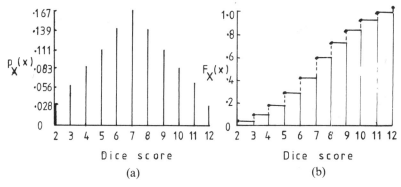

Fig. 2.6 pmf and cdf for example 2.5

THE PROBABILITY DENSITY FUNCTION (**pdf**)

A continuous random variable is a variable that is capable of achieving any value from an infinite number of values.

If we apply the frequenistic approach described in chapter 1 we can obtain an expression for, P, the probability of the variable taking on a particular value:

$$P = \frac{1}{N}$$

Unfortunately this approach will give $P = 0$ as N, the number of possible values, is infinite. We cannot, therefore, think in terms of the probabilities of occurrence of single values when we wish to obtain an expression for the probability density function of a continuous random variable. We must think in terms of ranges of values.

For a continuous random variable, X, we can consider that its full range of possible values is divided into infinitesimal lengths, each of size dx. With this approach the probability density function of X can be defined as function, $f_X(x)$, where $\int f_X(x)\, dx$ is the sum of the probabilities over the range dx.

$\int f_X(x)\, dx$ is the value of probability that X will take on a value within the range x to $(x + dx)$.

Remembering that $f_X(x)$ is a sum of probabilities, not a single probability, it will be seen that it can therefore have a value greater than 0, i.e.

$$f_X(x) \geq 0$$

Values in different ranges are mutually exclusive which means that, for a range of possible values of X, varying from a to b, the probability that X will lie within this range is the probability that $a \leq X \leq b$, i.e. that X must have a value a, or b, or somewhere in between. That is:

$$\int_a^b f_X(x)\, dx = P[a \leq X \leq b]$$

Obviously if a and b are the minimum and maximum values then the area under the function must be 1.0, i.e.

$$\int_{-\infty}^{\infty} f_X(x)\, dx = 1$$

(Note the convention of symbolising this total range as between $-$ and $+$ infinity.)

Two conditions that must therefore be satisfied by a probability density function are:

(1) $$f_X(x) = 0$$

(2) $$\int_{-\infty}^{\infty} f_X(x)\, dx = 1$$

Example 2.6

Prepare a probability density function that will fit the observed data of example 2.3.

Solution
The first step is to choose a suitable histogram and its associated frequency polygon.

It has been illustrated that a large variety of differently shaped histograms can be obtained from the same data and, as each one will have its separate pdf, a certain amount of judicious selection is required if the pdf obtained is to be a realistic model of the measurements.

For this solution the histogram of Fig. 2.3c and the corresponding frequency polygon of Fig. 2.4c have been chosen. Figures 2.3c and 2.4c are shown superimposed on each other in Fig. 2.7a. A pdf which fits the data and is mathematically simple is shown as a dashed line in Fig. 2.7a and is reproduced in Fig. 2.7b. It consists of three straight lines extending over a range of values from 22.85 to 41.25 kN/m².

It is, of course, necessary to determine the value of the vertical dimension 'k' and this is obtained by the application of the condition that the area under the total pdf must equal 1.0, i.e.

$$0.5(30.9 - 22.85)k + (34.35 - 30.9)k + 0.5(41.25 - 34.35)k = 1.0$$

Hence:

$$k = 0.09153$$

Now $f_X(x)$ is the equation of the pdf which, in this case, consists of three straight lines. Hence, $f_X(x)$ must consist of three separate parts, over the ranges:

$$22.85 \leqslant X \leqslant 30.90$$
$$30.90 \leqslant X \leqslant 34.35$$
$$34.35 \leqslant X \leqslant 41.25$$

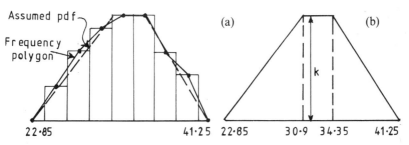

Fig. 2.7   Assumed pdf for example 2.6

For range $22.85 \leqslant X \leqslant 30.90$.

Slope of line $= k/(30.9 - 22.85) = 0.09153/8.05 = 0.01137$

Hence:

$$f_X(x) = 0.01137(X - 22.85)$$

For range $30.9 \leqslant X \leqslant 34.35$

$$f_X(x) = 0.09153$$

For range $34.5 \leqslant X \leqslant 41.25$

Slope of line $= 0.09153/(34.5 - 41.25) = -0.01356$

Hence:

$$f_X(x) = 0.09153 - 0.01356(X - 34.5)$$
$$= 0.55935 - 0.0136X$$

The probability density function decided upon is therefore:

$$f_X(x) = \begin{cases} 0.0114X - 0.2598 \\ 0.0915 \\ 0.5594 - 0.0136X \end{cases} \text{for} \begin{cases} 22.85 \leqslant X \leqslant 30.90 \\ 30.90 \leqslant X \leqslant 34.35 \\ 34.35 \leqslant X \leqslant 41.25 \end{cases}$$

CUMULATIVE DISTRIBUTION FUNCTION FOR A CONTINUOUS RANDOM VARIABLE

As for a discrete variable the cdf for a continuous random variable is expressed as:

$$F_X(x) = P[X \leqslant x]$$

The full range of a continuous variable can be considered as being from $-\infty$ to $+\infty$. Hence:

$$F_X(x) = P[-\infty \leqslant X \leqslant x]$$

or:

$$F_X(x) = \int_{-\infty}^{x} f_X(x)\,dx$$

The right hand expression can be confusing in that the symbol 'x' has been used to indicate both the variable of integration and the upper limit of the integration. The convention generally adopted is to use a dummy variable of integration, u, leading to the expression:

$$F_X(x) = \int_{-\infty}^{x} f_X(u)\,du$$

RELATIONSHIP BETWEEN THE PROBABILITY FUNCTION AND ITS CUMULATIVE DISTRIBUTION

$$\frac{dF_X(x)}{dx} = \frac{d}{dx}\int_{-\infty}^{x} f_X(u)\,du = f_X(x)$$

The value of the probability density function at x equals the slope of the cumulative distribution curve at x.

Example 2.7

The probability function of a continuous random variable, X, is given by the equation:

$$f_X(x) = 0.0064x(25 - x^2) \quad \text{for} \quad 0 \leqslant X \leqslant 5$$

Draw plots of both the pdf and the cdf of X.

Solution

$$F_X(x) = \int_{-\infty}^{x} f(u)\,du = \int_{0}^{x} 0.0064u(25 - u^2)\,du$$

$$= 0.08x^2 - 0.0016x^4$$

The two plots can be prepared by selecting suitable values for X:

x	0	1	2	3	4	4.5	5
$f_X(x)$	0	0.1536	0.2688	0.3072	0.2304	0.1368	0
$F_X(x)$	0	0.0784	0.2944	0.5904	0.8704	0.9639	1.0

The plots of $f_X(x)$ are shown in Fig. 2.8.

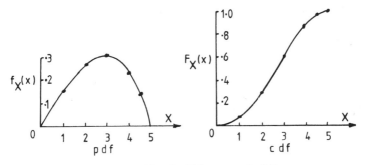

Fig. 2.8   pdf and cdf for example 2.7

Example 2.8

Using the pdf derived in example 2.6 prepare a plot of the cumulative distribution function.

Solution

$$F_X(x) = \int_{-\infty}^{x} f_X(u)\,du = \int_{22.85}^{30.9} f_X(u)\,du$$

$$+ \int_{30.9}^{34.35} f_X(u)\,du + \int_{34.35}^{41.25} f_X(u)\,du$$

For range $22.85 \leqslant X \leqslant 30.9$

$$f_X(x) = 0.0114X - 0.2598$$

$$F_X(x) = \int_{22.85}^{x} (0.0114u - 0.2598) \, du$$

$$= 0.0057x^2 - 0.2598x + 2.9603$$

i.e. when $X = 30.9$ then $F_X(x) = 0.3749$.
  For range $30.9 \leqslant X \leqslant 34.35$

$$f_X(x) = 0.0915$$

$$F_X(x) = 0.3749 + \int_{30.9}^{x} 0.0915 \, du = 0.3749 + 0.0915x - 2.8274$$

$$= 0.0915x - 2.4525$$

i.e. when $X = 34.35$, $F_X(x) = 0.6905$.
  For range $34.35 \leqslant X \leqslant 41.25$

$$f_X(x) = 0.5594 - 0.0136X$$

$$F_X(x) = 0.6905 + \int_{34.35}^{x} (0.5594 - 0.0136u) \, du$$

$$= 0.6905 + 0.5594x - 0.0068x^2 - 11.1919$$

$$= 0.5594x - 0.0068x^2 - 10.5014$$

i.e. when $X = 41.25$, $F_X(x) = 1.0$.

$F_X(x)$ values are obtained by selecting suitable values for X:

x	22.85	27.0	30.9	34.35	38.0	41.25
$F_X(x)$	0	0.1010	0.3749	0.6905	0.9366	1.0

The cumulative distribution function is shown in Fig. 2.9.

Expected value of a random variable

A random variable, X, is called random because it is never possible to say exactly what its value will be. If the equation of the pmf (or pdf) is known then its central value can be obtained. This value is known as the

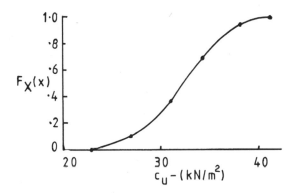

Fig. 2.9 cdf for example 2.8

expected value of X, or the expectation of X, and is given the symbol $E[X]$.

Whenever possible the expected values of random variables should be used in design calculations but there are many occasions when the pdf, or pmf, equations are not known and it becomes necesary to substitute $E[X]$ with the arithmetical mean value of a set of measurements of the variable, μ, μ_X, $\bar{x}$, m, or m_X.

It is seen therefore that $E[X]$ can generally be taken as equal to m_X, the mean of a sample of X, although it should be appreciated that there is a difference between the two terms.

If the pmf, or pdf, equation is known then the value of $E[X]$ can be determined by computation. This value is exactly equal to m_X, the population mean.

If the mean value of the variable has been estimated from a set of measurements or observations then this value should really be called the sample mean, m_s. There is little error involved in assuming that $m_X = m_s$ for a sample containing 30 values or more but for smaller samples there can be a significant difference between the two values.

For a discrete random variable, X, with a range of possible values from x_1 to x_n:

$$E[X] = \sum_{i=1}^{n} x_i P[X = x_i] = \sum x p_X(x)$$

Example 2.9

A discrete random variable x, has the following pmf:

$$p_X(x) = \begin{cases} 0.05 \\ 0.1 \\ 0.2 \\ 0.2 \\ 0.3 \\ 0.1 \\ 0.05 \end{cases} \quad \text{for} \quad X = \begin{cases} 1000 \\ 2000 \\ 3000 \\ 4000 \\ 5000 \\ 6000 \\ 7000 \end{cases}$$

Determine its expectation.

Solution

The expectation or the expected value equals $E[X]$

$$E[X] = 0.05 \times 1000 + 0.1 \times 2000 + 0.2 \times 3000 + \dots + 0.05 \times 7000$$

$$= 4100$$

Note: In the special case of all the probabilities of X being equal, the expression for $E[X]$, for a discrete variable, reduces to:

$$E[X] = \frac{x_1 + x_2 + x_3 + \dots + x_n}{n}$$

which is m_X, the arithmetic mean of the values $x_1, x_2, x_3, \dots, x_n$.

For a continuous variable, the analogous expression for $E[X]$ is:

$$E[X] = \int_{-\infty}^{\infty} x f_X(x) \, dx$$

Example 2.10

The pdf of a continuous random variable X is given by:

$$f_X(x) = \begin{cases} \frac{1}{72}x^2 & 0 \leqslant x \leqslant 6 \\ 0 & \text{elsewhere} \end{cases}$$

Determine the value of $E[X]$.

Solution

$$E[X] = \int_{-\infty}^{\infty} x f_X(x) \, dx = \int_{0}^{6} x(\tfrac{1}{72}x^2) \, dx$$

$$= [\tfrac{1}{288}x^4]_0^6 = 4.5$$

An approximate solution of the above problem can be obtained by graphical means and this approach may prove more convincing to some readers.

The plot of $f_X(x)$ is obtained by selecting a suitable set of values for X and is shown in Fig. 2.10.

X	0	1	2	3	4	5	6
$f_X(x)$	0	0.014	0.056	0.125	0.222	0.347	0.5

In a similar way to the preparation of a histogram, the range of possible X values is split into a suitable number of equal ranges, a total of six was used in this example.

For each particular range, approximate values for $f_X(x)$ and for the central value of X can be tabulated:

Range of X	Min. $f_X(x)$	Max. $f_X(x)$	Central X value	Average $f_X(x)$	$X . f_X(x)$
0–1	0	0.014	0.5	$0.5 \times 0.014 = 0.007$	0.004
1–2	0.014	0.056	1.5	$0.5(0.056 + 0.014) = 0.035$	0.053
2–3	0.056	0.125	2.5	0.091	0.228
3–4	0.125	0.222	3.5	0.174	0.609
4–5	0.222	0.347	4.5	0.285	1.286
5–6	0.357	0.5	5.5	0.429	2.360
					$\Sigma\ 4.477$

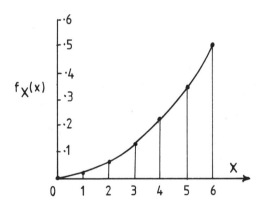

Fig. 2.10   Graphical solution for example 2.10

SOME EXPECTATION PROPERTIES

If a and b are constants then:

(i) $E[a] = a$
(ii) $E[aX] = aE[X]$
(iii) $E[a + bX] = a + bE[X]$

Expectations of functions of random variables

If X is an independent random variable and if Y is a function of X such that $Y = g(X)$, then Y is a random variable dependent on X. If X is a discrete variable then the expectation of Y is the summation of $g(X)$ multiplied by the probability of x_i, i.e.

$$E[Y] = \sum g(X)p_X(x)$$

If X is a continuous variable then:

$$E[Y] = \int_{-\infty}^{\infty} g(X)f_X(x)\, dx$$

Example 2.11

(a) Consider the discrete variable, X, of example 2.9 and assume $Y = X^2$. Then:

$$E[Y] = E[X^2] = \sum x^2 p_X(x)$$

$$= 0.05(1000)^2 + 0.1(2000)^2 + 0.2(3000)^2 + \ldots + 0.05(7000)^2$$

$$= 19\ 000\ 000$$

(b) If the variable X had been the continuous variable of example 2.10, then the expression of $Y = X^2$ would have been:

$$E[Y] = \int_0^6 x^2(\tfrac{1}{72}x^2)\, dx = [\tfrac{1}{360}x^5]_0^6 = 21.6$$

Variance of a random variable

The actual value, x, attained by a random variable, X, will generally not be equal to $E[X]$ and the difference between the two values, $(x - E[X])$, is known as the variation of x.

As $E[X]$ is equal to the population mean, m_X, the variation of x is also equal to $(x - m_X)$.

For any particular value of X the variation will be either negative, positive or zero. To remove the problem of negative numbers, the square of the variation is considered, $(x - m_X)^2$.

The average value of the squares of the variations over the whole range of possible X values is called the variance of X and given the symbol $\text{Var}(X)$. This average value is the expected value of $(X - m_X)^2$. Hence:

$$\text{Var}(X) = E[(X - m_X)^2]$$

Now, $(X - m_X)^2$ is a function of X and we can therefore write, for a discrete variable:

$$\text{Var}(X) = \text{E}[(X - m_X)^2] = \sum_{i=1}^{n} (x_i - m_X)^2 p_X(x)$$

and for a continuous variable:

$$\text{Var}(X) = \int_{-\infty}^{\infty} (x - m_X)^2 f_X(x)\,dx$$

Another expression for $\text{Var}(X)$ is obtained by expanding $\text{E}(X - m_X)^2$. As $\text{E}[X]$ is equal to m_X:

$$\text{E}(X - m_X)^2 = \text{E}[X^2] - 2\text{E}[X]m_X + m_X^2 = \text{E}[X^2] - \text{E}[X]^2$$

Hence, for a discrete variable:

$$\text{Var}(X) = \text{E}[X^2] - (\text{E}[X])^2$$
$$= x^2 p_X(x) - [x p_X(x)]^2$$

Example 2.12

A continuous random variable, X, has the following pdf:

$$f_X(x) = \begin{cases} 0.00025x & 80 \leqslant x \leqslant 120 \\ 0 & \text{elsewhere} \end{cases}$$

Determine the variance of X.

Solution

$$\text{E}[X] = \int_{80}^{120} x f_X(x)\,dx = \int_{80}^{120} x(0.00025x)\,dx = [0.00008333x^3]_{80}^{120}$$

$$= 101.33 = m_X$$

$$\text{Var}[X] = \text{E}[(X - m_X)^2] = \text{E}[(X - 101.33)^2]$$

$$= \int_{80}^{120} (x - 101.33)^2(0.00025x)\,dx$$

The value of this integral is 132.6 so that $\text{Var}(X)$ is equal to 132.6.

Standard deviation of a random variable

The square root of the variance is called the standard deviation and is given the symbol σ_X, or simply σ, when there is no risk of confusion. In mathematical symbols:

$$\sigma_X = \sqrt{\text{Var}[X]} = \sqrt{E(X - m_X)^2}$$

Obviously:

$$\text{Var}[X] = \sigma_X^2$$

The variance and the standard deviation of a variable are both measures of the amount of spread of its possible values from the mean value.

If a variable has values that tend to be close to the mean then the variance of the variable will be small whereas if the possible range of values extends for some distance either side of the mean then the variance will be large. (See Fig. 2.11.)

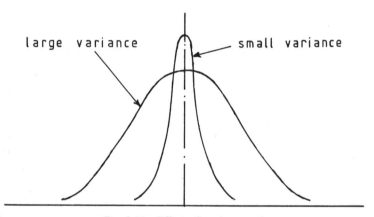

large variance   small variance

Fig. 2.11   Effect of variance value

Example 2.13

Determine the variance and standard deviation of the discrete random variable of example 2.9.

Solution

$$\text{Var}(X) = E[(X - m_X)^2] = \sum_{i=1}^{n} (x_1 - m_X)^2 p_X(x)$$

Now:

$$m_X = E[X] = 4100$$

$$\text{Var}(X) = 0.05(1000 - 4100)^2 + 0.1(2000 - 4100)^2 + 0.2(3000 - 4100)^2$$
$$+ 0.2(4000 - 4100)^2 + 0.3(5000 - 4100)^2 + 0.1(6000 - 4100)^2$$
$$+ 0.05(7000 - 4100)^2$$

$$= 2\,190\,000$$

$$\sigma_X = \sqrt{2\,190\,000} = 1479.9$$

Alternatively the other formula may be used:

x	$p_X(x)$	$xp_X(x)$	$x^2p_X(x)$
1000	0.05	50	50 000
2000	0.1	200	400 000
3000	0.2	600	1 800 000
4000	0.2	800	3 200 000
5000	0.3	1500	7 500 000
6000	0.1	600	3 600 000
7000	0.05	350	2 450 000
Total	1.00	4100	18 946 500

$$\mathrm{Var}(X) = x^2p_X(x) - [xp_X(x)]^2 = 18\,946\,500 - 4100^2 = 2\,190\,000$$

Mean and standard deviation of a statistical sample

In civil engineering the term 'sample' means a single item, such as a soil sample or a single concrete test cube whereas in statistics the term means a set of results or values.

Generally n soil samples will yield one statistical sample which consists of n values of the parameter measured with each soil sample. This statistical sample can be regarded as a discrete variable with $p_X(x)$ equal to $1/n$. The mean value, m_X, is given by:

$$m_X = \frac{x_1 + x_2 + x_3 + \ldots x_n}{n}$$

And:

$$\mathrm{Var}(X) = E[(X - m_X)^2] = \sum_{i=1}^{n} (x_i - m_X)^2 p_X(x) = \sum \frac{(x - m_X)^2}{n}$$

Hence, the standard deviation, s_X, is given by:

$$s_X = \sqrt{\frac{\sum (x - m_X)^2}{n}}$$

Example 2.14

Determine the mean and standard deviation of the values

$$6, 5, 8, 5, 5, 6, 7, 8, 4, 6$$

Solution
The mean value is given by:

$$\frac{6 + 5 + 8 + 5 + 5 + 6 + 7 + 8 + 4 + 6}{10} = 6$$

Now:

$$\text{Var} = \sum_{i=1}^{10} (x_i - m_X)^2$$

Hence:

$$s = \sqrt{\frac{(-1)^2 + 2^2 + (-1)^2 + (-1)^2 + 1 + 2^2 + (-2)^2}{10}} = 1.26$$

Note: The simplicity of the standard deviation formula for a set of values illustrates the simplicity of the arithmetic, but not the tedium, involved when the number of values is large. This tedium is greatly reduced if the formula is rearranged and written as

$$s_X = \sqrt{\frac{\sum x^2}{n} - m_X^2}$$

which leads to the same value of 1.26 for the standard deviation.

Standard deviation of a population – Bessel's correction

As illustrated in examples 2.12 and 2.13 the variance of a random variable can be determined exactly if its pmf or its pdf is known. Unfortunately, in civil engineering, the functions representing the pmfs and the pdfs of the relevant parameters can only be estimated and the engineer is obliged to use standard deviation values obtained from statistical samples, as illustrated in example 2.14.

However, a sample variance invariably underestimates the population variance simply because the value of the sample mean is not usually the same as the value of the population mean. Because of this, in many cases, the sum of the deviations of the values about the sample mean is smaller than the sum of the squares of the deviations about the population mean resulting in the sample variance inevitably being less than the population variance.

In order to estimate the standard deviation of the population, given s, the standard deviation value of a representative sample of the variable, we make use of Bessel's correction which allows for this underestimation, thus:

$$\hat{\sigma} = s\sqrt{\frac{n}{n-1}}$$

where n is the number of values in the sample and the circumflex indicates that the value has been estimated.

If, for instance, the values quoted in example 2.14 had been drawn from a large population then the best estimate for the standard deviation of the population would be:

$$1.26\sqrt{\tfrac{10}{9}} = 1.33$$

The estimated value of the population standard deviation can be obtained directly from the sample values by using the formula:

$$\hat{\sigma} = \sqrt{\frac{\Sigma\,(x - m_X)^2}{n - 1}}$$

Some operators only use the formula involving $n - 1$ and, as only small statistical samples are possible in civil engineering, the author recommends this practice.

For large values of n, $(n > 30)$, the difference between s and $\hat{\sigma}$ becomes negligible.

Dispersion of values

The variance (or standard deviation) of a variable is an indication of the amount of dispersion of the values that the variable can have.

If the variance is low then the values will be concentrated near to the mean whilst if the variance is high the values will be scattered within a wide range whose limits are some distance from the mean (Fig. 2.11).

For any variable the units of its standard deviation are the same as the units for its mean whereas the units of its variance are squared. Because of this most statistical work is carried out using standard deviations.

As the reader will appreciate, it is quite possible for a completely different set of data to have the same mean value as another set of data. A knowledge of the arithmetic mean is therefore very limited in application unless it is accompanied by the relevant frequency curve or some information such as the standard deviation.

Coefficient of variation

It is often useful to be able to compare the degree of spread of two different variables of different units, say one in kN/m^3 and one in kN. This is achieved by a term known as the coefficient of variation which is simply a dimensionless ratio created by dividing the standard deviation of the variable by its mean. The coefficient of variation, V, is given by:

$$V = \frac{\sigma}{m}$$

Example 2.15

Measured values of a variable X and of a variable Z are set out below.

Variable X:	20	25	30	35	40	45	50	55	60
Variable Z:	120	125	130	135	140	145	150	155	160

Which set of values has the greatest dispersion?

Solution

By inspection it can be seen that the two sets of values have the same value of standard deviation which works out to be 12.91. The mean value of variable X is 40 and the mean value of variable Z is 140.

$$V(X) = \frac{12.91}{40} = 32.3\% \quad \text{and} \quad V(Z) = \frac{12.91}{140} = 9.2\%$$

The spread of the values of variable X is much greater than that of the values of variable Z.

Other mean values

Although the arithmetic mean is the value most commonly used in civil engineering there are two other ways of expressing mean values and the reader should be aware of them.

THE HARMONIC MEAN

The harmonic mean of a set of values is simply the reciprocal of the mean sum of the reciprocals of the values. In symbols, if there are n values:

$$\text{Harmonic mean} = \frac{n}{\sum\limits_{i=1}^{n} \left(\frac{1}{x_i}\right)}$$

THE GEOMETRIC MEAN

The geometric mean of a set of n values, x_i, is the nth root of the product of the values. That is:

$$\text{Geometric mean} = \sqrt[n]{x_1 . x_2 . x_3 . x_4 \ldots x_N}$$

Other ways of expressing the average value

It must be remembered that the arithmetic mean takes the extreme values of the range into account. This can be a disadvantage when a sample contains only a few values and, for such cases, use can be made of the mode and the median values. The two terms are defined below.

THE MODE (OR MODAL VALUE)

The word 'mode' means 'fashion' so the modal value of a set of numbers is simply the value most in fashion, i.e. the value that occurs most often.

The mode of a discrete variable is therefore the value corresponding to the peak of its frequency curve (group size 6 in Fig. 2.1 for example) and, for a continuous variable, the central value of the highest block of its histogram (33.2 in Fig. 2.3c for example).

THE MEDIAN

The median is the central value of a set. On the one side all the values are less then the median, on the other side all the values are greater than the median.

In other words, if the values are placed in ascending order of magnitude, then the median is the central value (if the number of values is odd) or the mean of the two central values (if the number of values is even).

Example 2.16

Determine the arithmetic, the harmonic and the geometric means, the mode and the median for the following set of numbers.

$$2, 3, 3, 3, 5, 6, 7, 8, 10, 11, 14, 24$$

Solution

$$\text{Arithmetic mean} = \frac{2 + 3 + 3 + 3 + 5 + 6 + 7 + 8 + 10 + 11 + 14 + 24}{12}$$

$$= 8$$

Harmonic mean

$$= \frac{12}{\frac{1}{2} + \frac{1}{3} + \frac{1}{3} + \frac{1}{3} + \frac{1}{5} + \frac{1}{6} + \frac{1}{7} + \frac{1}{8} + \frac{1}{10} + \frac{1}{11} + \frac{1}{14} + \frac{1}{24}}$$

$$= 4.92$$

Geometric mean

$$= \sqrt[12]{2 \times 3 \times 3 \times 3 \times 5 \times 6 \times 7 \times 8 \times 10 \times 11 \times 14 \times 24}$$

$$= 6.22$$

mode = most common number = 3 .

Median = central number = mean of 6 and 7 = 6.5

Skewness

Up to this point it has been tacitly assumed that any distribution discussed in this chapter is symmerical with the mean, the median and the mode all coinciding at the central value of the distribution. However, this will not always be the case. Many distributions are asymmetrical, i.e. non-symmetrical with one tail of the distribution longer than the other. Because of this, we introduce the term skewness or, more correctly, the coefficient of skewness, as a measure of the amount of asymmetry. If the longer tail is to the right of the mean then the distribution has a right-hand skew whereas if the longer tail is to the left then the skew is left handed.

A symmetrical distribution has no asymmetry and therefore its skewness is zero. The convention used for asymmetry is that the skewness of a right-hand skew is positive whereas for a left-hand skew the skewness is negative, see Fig. 2.12.

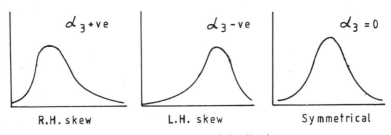

Fig. 2.12   Skewness of distributions

For moderately skewed distributions, either left or right, it can be shown that there is an approximate relationship between the arithmetic mean, the mode and the median:

$$\text{mode} = \text{mean} - 3(\text{mean} - \text{median})$$

The extreme values of a distribution have a much greater effect on the mean value than on the mode or the median values so that the mean value of an asymmetrical distribution is always further into the longer tail of the distribution than the other two values.

A measure of skewness is therefore (mean − mode) which will be negative for left-handed skews and positive for right-handed ones. In order to remove the dimensions we simply divide by the standard deviation of the distribution to give the formula:

$$\text{Coefficient of skewness} = \frac{\text{mean} - \text{mode}}{\text{standard deviation}}$$

or:

$$\alpha_3 = \frac{3(\text{mean} - \text{median})}{\text{standard deviation}}$$

Moments

It is often convenient to think in terms of the moments of area of the pdf (or pmf) diagram about some point.

If the moment is about the origin then it is known as an ordinary moment and is given the symbol μ'_r. For a discrete variable the rth moment about the origin is given by:

$$\mu'_r = \sum x^r p_X(x) = E[X^r]$$

and for a continuous variable:

$$\mu'_r = \int x^r f_X(x)\, dx = E[X^r]$$

As will be seen later in this section, we are generally only interested in the first four moments. Expressions for these moments, for a continuous variable, are:

$$\mu'_1 = E[X] = \int x f_X(x)\, dx$$

$$\mu'_2 = E[X^2] = \int x^2 f_X(x)\, dx$$

$$\mu'_3 = E[X^3] = \int x^3 f_X(x)\, dx$$

$$\mu'_4 = E[X^4] = \int x^4 f_X(x)\, dx$$

The most important of the ordinary moments is the first moment:

$$\mu_1' = E[X^1] = m$$

Obviously moments can be taken about any point but the only other meaningful point is the mean value. Moments of area about the mean are known as mean or central moments and given the symbol μ_r. For a discrete variable:

$$\mu_r = \sum (x - m)^r f_X(x) = E[(X - m)^r]$$

and for a continuous variable:

$$\mu_r = \int (x - m)^r f_X(x)\,dx = E[(X - m)^r]$$

Note that μ_1 is equal to $E[X - m]$ and must therefore be equal to zero.

For a continuous variable the various expressions for higher central moments are:

$$\mu_2 = E[(X - m)^2] = \int (x - m)^2 f_X(x)\,dx$$

$$\mu_3 = E[(X - m)^3] = \int (x - m)^3 f_X(x)\,dx \ldots \text{etc.}$$

Ths most important of the central moments is the second moment:

$$\mu_2 = E[(X - E(X))^2] = \text{Var}[(X]$$

Relationship between ordinary and central moments

It can be seen from the above formulae that the ordinary moments are much simpler to evaluate directly than the central moments. Fortunately, a relationship exists between the two sets of moments so that μ_r values can be expressed in terms of μ_r' values.

$$\mu_2 = \mu_2' - m^2$$
$$\mu_3 = \mu_3' - 3\mu_2'm + 2m^3$$
$$\mu_4 = \mu_4' - 4\mu_3'm + 6\mu_2'm^2 - 3m^4$$

Example 2.17

The probability density function of a random variable, X, is:

$$f_X(x) = \tfrac{5}{48}(8x - x^4) \qquad 0 \leqslant x \leqslant 2$$

Determine the first three moments of its pdf, (i) about the origin, and (ii) about its mean value.

Solution

(i)
$$\mu_1' = E[X] = \int xf_X(x)\,dx = \int_0^2 \frac{5}{48} x(8x - x^4)$$

$$= \frac{5}{48}\left[8x^3 - \frac{x^6}{6}\right]_0^2 = 1.111$$

$$\mu_2' = E[X^2] = \frac{5}{48}\left[2x^4 - \frac{x^7}{7}\right]_0^2 = 1.429$$

$$\mu_3' = E[X^3] = \frac{5}{48}\left[\frac{8x^5}{5} - \frac{x^8}{8}\right]_0^2 = 2.000$$

(ii)
$$\mu_1 = 0$$

$$\mu_2 = \mu_2' - m^2 = 1.429 - 1.111^2$$
$$= 0.195$$

$$\mu_3 = \mu_3' - 3\mu_2'm + 2m^3$$
$$= 2.000 - 3 \times 1.429 \times 1.111 + 2 \times 1.111^3$$
$$= -0.020$$

Note: The reader might like to check the relationship formulae by evaluating μ_2 and μ_3 directly, although the work involved becomes tedious as the power of the moment increases. For example:

$$\mu_2 = \int (X - m)^2 f_X(x)\,dx = \int_0^2 (x - 1.111)^2 \tfrac{5}{48}(8x - x^4)dx$$

$$= 0.195$$

$$= \tfrac{5}{48}\int_0^2 (-x^6 + 2.222x^5 - 1.2345x^4 + 8x^3 - 17.776x^2 + 9.876x)\,dx$$

$$= 0.195$$

The coefficient of skewness α_3

An approximation for α_3, which can be used for a set of data, has already been given. If the pmf or pdf is known, then α_3 can be found from an expression involving the third central moment:

$$\text{Coefficient of skewness} = \alpha_3 = \frac{\mu_3}{\sigma^3} = \frac{E[X - m^3]}{\sigma^3}$$

α_3 will be zero if the distribution is symmetrical. It will have a positive value if there is a right-hand skew and negative if there is a left-hand skew.

Kurtosis

When most of the values of a distribution are concentrated around the mean value then the distribution has a pronounced peak whereas it is much flatter when the values are more evenly spread throughout the range (Fig. 2.11).

The degree of peakedness of a distribution is measured by the dimensionless coefficient of kurtosis, α_4, which involves the fourth central moment:

$$\text{Kurtosis} = \alpha_4 = \frac{\mu_4}{\sigma^4} = \frac{E[(X - m)^4]}{\sigma^4}$$

The peakedness of a particular distribution is assessed by examining just how much its kurtosis is greater, or smaller, than 3.0, which is the value of kurtosis for the normal distribution (described in chapter 3).

Exercises

HISTOGRAM

2.1 The heights, in centimetres, of 70 male students are recorded below.

172	173	170	176	168	179	175
173	181	177	186	178	177	172
181	170	164	172	176	167	179
167	173	176	183	176	187	170
175	172	173	159	171	168	177
179	169	174	179	175	169	162
172	169	181	171	166	174	178
176	175	165	185	177	188	174
153	168	176	176	177	171	187
178	173	175	170	177	171	183

Using 8 cells of equal width draw the histogram, the frequency polygon, the cumulative frequency curve and the relative cumulative frequency curve.

Hint: to have 8 equal cell widths it is simplest to extend range of heights from 150 to 190 cm.

Answer: Minimum value = 153 cm; maximum value = 188 cm.

(1) Cell	(2) Frequency of values	(3) Cumulative frequency	(4) Relative cumulative frequency
150–155	1	1	0.014
>155–160	1	2	0.029
>160–165	3	5	0.071
>165–170	15	20	0.286
>170–175	21	41	0.586
>175–180	19	60	0.857
>180–185	6	66	0.943
>185–190	4	70	1.000

Column (1) is used in the preparation of both the histogram and the frequency polygon.

PROBABILITY FUNCTIONS

2.2   A fair coin is tossed four times and X is the number of heads obtained in these four throws.

Obviously X is a discrete random variable with a range of integer values from 0 to 4.

Prepare a line diagram that represents $p_X(x)$, the probability mass function of X.

Hint:

Probability of a head with each throw, $P[H] = \frac{1}{2}$.

Probability of a tail with each throw, $P[T] = \frac{1}{2}$.

Probability of 3 heads $= P[H].P[H].P[H].P[T] = \frac{1}{16}$ but three heads can be scored in four different ways.

Hence, total probability of three heads $= \dfrac{4 \times 1}{16} = \frac{1}{4}$.

Answer:

Horizontal plot (x)	0	1	2	3	4
Vertical plot $(p_X(x))$	$\frac{1}{16}$	$\frac{1}{4}$	$\frac{3}{8}$	$\frac{1}{4}$	$\frac{1}{16}$

EXPECTATION AND VARIANCE

2.3   Show that, if:

$$f_X(x) = k\left(x + \frac{x^2}{10}\right) \qquad \text{for } 0 \leqslant X \leqslant 9,$$

then $k = 0.0154$.

Determine the expected value and the variance of X.

Answer: $E[X] = 6.268$; $Var[X] = 0.015$.

2.4   The pmf for a discrete random variable, X, is set out below.

$$p_X(x) = \begin{cases} 0.03 \\ 0.15 \\ 0.22 \\ 0.25 \\ 0.20 \\ 0.10 \\ 0.05 \end{cases} \qquad \text{for } X = \begin{cases} 0 \\ 20 \\ 40 \\ 60 \\ 80 \\ 100 \\ 120 \end{cases}$$

Determine the expected value and the variance of X.

Answer: $E[X] = 58.8$; $Var[X] = 854.56$.

MEANS AND STANDARD DEVIATION

2.5 (i) Determine the arithmetic mean, the median and the standard deviation for the following values.

24, 20, 16, 25, 18, 30, 28, 15, 22, 32, 28, 19

(ii) Estimate a value for the standard deviation of the population.

Answer: (i) 23.08; 23; 5.39.   (ii) 5.63.

2.6 (i) Considering the first four columns of the values listed in exercise 2.1 determine the arithmetic mean, the geometric mean, the harmonic mean, the variance and standard deviation of these values.

(ii) Determine the approximate value of the standard deviation of the population from which the sample was chosen.

Answer:

Arithmetic mean = 173.42   Variance = 41.99 (sample)
Geometric mean = 1.25   Standard deviation = 6.40 (sample)
Harmonic mean = 173.18   Standard deviation = 6.48 (population)

MOMENTS

2.7 A continuous random variable, X, has the following pdf:

$$f_X(x) = 0.0494x^3 \qquad 0 \leqslant x \leqslant 3$$

Determine the first 4 moments of its pdf about: (i) the origin; (ii) the mean value of X.

Answer: (i) 2.40; 6.00; 15.43; 40.51. (ii) 0; 0.24; -0.122; 0.209.

Chapter Three

Common Probability Distributions

This chapter describes the three most important probability distributions used for civil engineering situations.

The first two, the binomial and the Poisson distributions, are used for discrete variables and the last one, the normal distribution, is the most important of the continuous distributions.

Some introductory material must be introduced before the distributions can be examined.

PERMUTATIONS

Let us suppose that we have a set of n foreign coins, each one different from the others, and let us suppose that we decide to arrange r of these coins in a row. There are obviously a large number of possible arrangements. Every arrangement will be different as each depends on the variation of the order in which the coins are placed. Each single possible arrangement is known as a permutation.

With n coins there are n ways of selecting the first one, $(n - 1)$ ways of selecting the second one, $(n - 2)$ ways of selecting the third one and $(n - r + 1)$ ways of selecting the rth (i.e. the final) one.

The total number of possible arrangements of r coins out of n, i.e. the total number of permutations, is represented by the symbol $_nP_r$, and must be the product of all these ways. Hence:

$$_nP_r = n(n - 1)(n - 2)(n - 3)\ldots(n - r + 2)(n - r + 1)$$

Note the convention used for the symbol $_nP_r$. The number placed in front of P is n, the total number of objects, whereas the number after P is r, the number of the objects that are to be selected.

If all the coins are to be placed in the row then we have the unique case of r equalling n and $_nP_r$ becomes $_nP_n$, given by:

$$_nP_n = n(n - 1)(n - 2)\ldots(n - n + 2)(n - n + 1) = n!$$

($n!$ is called factorial n). It can be shown that the expression for $_nP_r$ can be written as:

$$_nP_r = \frac{n!}{(n-r)!}$$

and it is seen that, when r equals n, then:

$$_nP_n = \frac{n!}{0!} = n!$$

with the assumption that 0! is 1.

Example 3.1

If there were twenty coins how many different arrangements of five coins would be possible?

Solution

$$_nP_r = {_{20}P_5} = \frac{20!}{(20-5)!} = \frac{20!}{15!} = 1\,860\,480$$

TREATMENT OF A SET THAT CONTAIN SUBSETS

Quite often the objects from which a selection is to be made are not all different from each other. It may be that a set of n objects has n_1 objects all the same, n_2 objects all the same ..., n_j objects all the same such that:

$$n = n_1 + n_2 + \ldots + n_j$$

If all the objects are to be selected then the number of permutations is $_nP_{n_1,n_2,n_3,\ldots,n_j}$ where:

$$_nP_{n_1,n_2,n_3,\ldots,n_j} = \frac{n!}{n_1!\,n_2!\,n_3!\ldots n_j!}$$

Example 3.2

Determine the number of different arrangements that can be made of the letters that occur in 'GEOTECHNICS'.

Solution
G, O, T, H, N, I and S occur once, C and E occur twice. Hence total number of permutations is given by:

$$\frac{11!}{2!\,2!\,1!} = 9\,979\,200$$

(If all eleven letters had been different, the number of permutations would have been almost 40 million.)

COMBINATIONS

A complete set of permutations includes every arrangement possble. For example an arrangement X_1, X_3, X_2, X_4 is considered different from the arrangement X_1, X_2, X_3, X_4.

A combination simply considers the objects in a group, without any regard to their arrangement. The combination X_1, X_3, X_2, X_4 is the same as X_1, X_2, X_3, X_4, exactly the same as for a set.

The total number of sets of r objects that can be selected from n objects is given the symbol $_nC_r$ (the total possible number of combinations of n things taken r at a time). $_nC_r$ is often written as:

$$\binom{n}{r}$$

and equals:

$$\frac{_nP_r}{r!} = \frac{n!}{r!\,(n-r)!}$$

Example 3.3

How many groups of five different coins can be created from a set of 20 different coins?

Solution

$$_{20}C_5 = \frac{_{20}P_5}{5!} = \frac{20!}{(20-5)!\,5!} = 15\ 504$$

Note: The idea of combinations has already been encountered in exercise 2.2 at the end of the previous chapter – the number of ways it is possible to obtain three heads with four throws of a fair coin could be found by enumeration and was stated to be four.

Now, from the foregoing, it is seen that this number is the combination of three out of four, i.e.

$$_4C_3 = \frac{4!}{(4-3)!\,3!} = 4$$

Binomial coefficients

The symbol $\binom{n}{r}$ is often referred to as a binomial coefficient as the binomial expansion for $(x+y)^n$ can be written as:

$$(x+y)^n = \binom{n}{r}x^{n-r}y^r$$

where n is a positive integer and r is an integer varying in steps of 1.0 from 0 to n.

When written out in full the expression is:

$$(x + y)^n = x^n + \binom{n}{1}x^{n-1}y + \binom{n}{2}x^{n-2}y^2 + \ldots + y^n$$

Note that:

$$\binom{n}{0} = \binom{n}{n} = 1.0$$

Example 3.4

Expand $(x + y)^3$.

Solution

$$(x + y)^3 = x^3 + \binom{3}{1}x^{3-1}y + \binom{3}{2}x^{3-2}y^2 + y^3$$

$$= x^3 + 3x^2y + 3xy^2 + y^3$$

Example 3.5

A fair die is to be thrown eight times. Determine the probabilities of: (i) obtaining three 3s and (ii) obtaining no 3s.

Solution

Let $p = P[3] =$ the probability of a 3
Let $q = P[\bar{3}] =$ the probability of no 3s
Then, $p = \frac{1}{6}$ and $q = (1 - p) = \frac{5}{6}$

A possible combination for the occurrence of three 3s is:

$$3 \quad 3 \quad 3 \quad \bar{3} \quad \bar{3} \quad \bar{3} \quad \bar{3} \quad \bar{3}$$

and the probability of this event is:

$$\tfrac{1}{6} \cdot \tfrac{1}{6} \cdot \tfrac{1}{6} \cdot \tfrac{5}{6} \cdot \tfrac{5}{6} \cdot \tfrac{5}{6} \cdot \tfrac{5}{6} \cdot \tfrac{5}{6} = 1.86 \times 10^{-3}$$

But, there are $_8C_3$ (56) combinations that can cause three 3s to appear. Hence, the total probability is $56 \times 1.86 \times 10^{-3} = 0.104$.

Note: From the above example we can write down a general expression for the probability of obtaining r 3s from a total of n throws:

$$\text{Probability} = {_nC_r}p^r q^{(n-r)} = \binom{n}{r}p^r q^{(n-r)}$$

For part (i) $r = 3$:

$$\text{Probability} = \frac{8!}{5!\,3!} \, (\tfrac{1}{6})^3 \cdot (\tfrac{5}{6})^5 = 0.104$$

For part (ii) $r = 0$:

$$\text{Probability} = {}_8C_0 p^0 q^8 = \frac{8!}{8!} \, q^8 = q^8 = (\tfrac{5}{6})^8 = 0.233$$

The Binomial distribution

Example 3.5 is an illustration of this important distribution. In a series of independent repeatable experiments, such as the tossing of coins or dice, the actual throw is referred to as a trial and the score obtained is termed an event. Independent repeatable trials are often referred to as Bernoulli trials.

For every trial there is a probability, p, related to the particular event, e.g. a head or a tail with a coin, a 3 with a die, etc. and there is also a probability of its non-occurrence, q, equal to $(1 - p)$.

It is often convenient to refer to p, the probability that the event will occur in a single Bernoulli trial, as the probability of success and q as the probability of failure.

$p_X(x)$, the probability that the event will happen exactly x times out of n Bernoulli trials, i.e. the probability of x successes and $(n - x)$ failures, can be written as the probability mass function:

$$p_X(x) = P[X = x] = \binom{n}{x} p^x q^{(n-x)}$$

where X is the discrete variable representing the number of times the event can occur ($0 \leqslant X \leqslant n$).

This discrete probability mass function is called the binomial distribution because, if we successively insert values for X from 0 to n, the values obtained for the function are equal to the successive terms of the binomial expansion of $(q + p)^n$.

The shape of the binomial distribution, which, being discrete, is represented by a line diagram, varies considerably with the values of both n and p.

This is demonstrated by Fig. 3.1 depicting two binomial distributions, both with n equal to 10 but with $p = 0.2$ in the left-hand graph and $p = 0.6$ in the right-hand one.

Note: The notation B(n, p) is used to indicate the binomial distribution that has the parameters n and p.

For example, the two distributions shown in Fig. 3.1 would be represented numerically by the symbols B(10, 0.2) and B(10, 0.6) respectively.

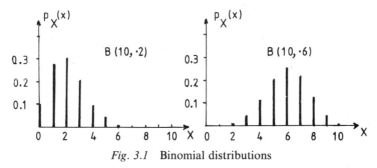

Fig. 3.1   Binomial distributions

Example 3.6

It is estimated that 10% of the steel bolts produced at a factory are defective. Tests, carried out at periods during production, consist of the random selection of groups of five bolts for inspection. If X is the number of faulty bolts found in a sample group of five bolts determine its probability mass function, mean value and standard deviation.

Solution
The probability of success, p (i.e. of finding a faulty bolt) = 0.10
The probability of failure, q (i.e. of not finding a faulty bolt) = 0.90

The possible values of X extend from zero to five and the pmf of X is therefore:

$$p_X(x) = \binom{5}{x} p^x q^{(5-x)}$$

Values of $p_X(x)$ can now be obtained for all values of X and are tabulated below along with the values of $x p_X(x)$ and $x^2 p_X(x)$.

X	$p_X(x)$	$x p_X(x)$	$x^2 p_X(x)$
0	0.59049	0.0	0.0
1	0.32805	0.32805	0.32805
2	0.07290	0.14580	0.29160
3	0.00810	0.02430	0.07290
4	0.00045	0.00180	0.00720
5	0.00001	0.00005	0.00025
Totals	1.00000	0.5	0.7

Now:

$$m_X = E[X] = \sum x p_X(x) = 0.5$$

$$\text{Var}(X) = \sum x^2 p_X(x) - \left(\sum x p_X(x)\right)^2 = 0.7 - 0.5^2 = 0.45$$

Hence:

$$\sigma_X = \sqrt{0.45} = 0.67$$

Note: There is actually no need to go through the above procedure in order to determine m_X and σ_X as there are formulae for both of them. For a binomial distribution:

$$m_X = np$$

$$\sigma_X = \sqrt{npq}$$

If the reader uses these formulae he will obtain the same values for m_X and σ_X as obtained by the procedure used in the example.

Example 3.7

X is a discrete variable with the probability distribution $B(18, 0.35)$. Determine the value of $F_X(4)$.

Solution

$$F_X(4) = P[X \leqslant 4]$$

For a binomially distributed variable:

$$P[X \leqslant 4] = p_X(0) + p_X(1) + p_X(2) + p_X(3) + p_X(4)$$

$$p_X(0) = P[X = 0] = \binom{18}{0} 0.35^0 0.65^{18} = 0.0004$$

$$p_X(1) = P[X = 1] = \binom{18}{1} 0.35^1 0.65^{17} = 0.0042$$

$$p_X(2) = P[X = 2] = \binom{18}{2} 0.35^2 0.65^{16} = 0.0190$$

$$p_X(3) = P[X = 3] = \binom{18}{3} 0.35^3 0.65^{15} = 0.0547$$

$$p_X(4) = P[X = 4] = \binom{18}{4} 0.35^4 0.65^{14} = 0.1104$$

$$\text{Total} = 0.1886$$

Hence:

$$F_X(4) = 0.1886$$

EVALUATION OF cdf VALUES FOR A BINOMIAL DISTRIBUTION

Example 3.7 illustrates that, although individual values of $p_X(x)$ can be easily obtained from the formula for the pmf, it can be tedious to determine a particular cdf value. particularly if n is large and x is close to n.

Tabulated values are available and the writer has found that the tables prepared by Neave (1978) are more than adequate for the level of probability used in this text.

A part of the table that give values of the equation:

$$F_X(x) = \sum_{r=0}^{x} \binom{n}{p} p^r(1 - p)^{n-r}$$

and is appropriate to example 3.7 is reproduced below.

n	x	p	0.30	0.35	0.40
18	0		0.0016	0.0004	0.0001
	1		0.0142	0.0046	0.0013
	2		0.0600	0.0236	0.0082
	3		0.1646	0.0783	0.0328
	4		0.3327	0.1886	0.0942
	5		0.5344	0.3550	0.2088
	6		0.7217	0.5491	0.3743
	7		0.8593	0.7283	0.5634
	8		0.9404	0.8609	0.7368
	9		0.9790	0.9403	0.8653
	10		0.9939	0.9788	0.9424
	11		0.9986	0.9938	0.9797
	12		0.9997	0.9986	0.9942
	13		1.0000	0.9997	0.9987
	14		1.0000	1.0000	0.9998
	15		1.0000	1.0000	1.0000
	16		1.0000	1.0000	1.0000
	17		1.0000	1.0000	1.0000

The value of $F_X(4)$, required for example 3.7, can be read off directly from the table.

Tabulated cdf values are available for n values from 1 to 20. For larger values of n, it is possible to determine cdf values because of the fact that the binomial distribution approximates to another discrete probability distribution, the Poisson distribution, and also to a continuous distribution, the normal distribution, both of which are described below.

The Poisson distribution

In order to use the binomial distribution, it is necessary to know the number of 'successes' (the number of times the event happened) and the number of 'failures' (the number of times the event did not happen). Because of this, the most common application of the binomial distribution is in the field of testing, as illustrated in example 3.6.

However, there are many problems in civil engineering where, although the number of successes can be measured, it is not possible to determine the number of failures. For example, it is a simple enough matter to determine the number of transverse cracks that have occurred in a kilometre of carriageway but there is no way of determining, or giving meaning to, the number of cracks that have not occurred.

For such problems the Poisson distribution is used. Although this is still a discrete distribution, it only requires knowledge of the number of successes obtained in a given period.

The base of Napierian logarithms, e, is equal to the sum of an infinite series:

$$e = \frac{1}{0!} + \frac{1}{1!} + \frac{1}{2!} + \frac{1}{3!} + \frac{1}{4!} + \frac{1}{5!} + \cdots$$

(The sum of the first four terms equals 2.7183). It can be shown that e^k, or $\exp(k)$, is also the sum of an infinite series:

$$\exp(k) = \frac{k^0}{0!} + \frac{k^1}{1!} + \frac{k^2}{2!} + \frac{k^3}{3!} + \frac{k^4}{4!} + \frac{k^5}{5!} + \cdots$$

$$= 1 + k + \frac{k^2}{2!} + \frac{k^3}{3!} + \frac{k^4}{4!} + \frac{k^5}{5!} + \cdots$$

Now, if we multiply e^k by e^{-k} we obtain 1.0. Therefore the terms in the expression for $\exp(k)$ summate to 1.0 when they are multiplied by $\exp(-k)$:

$$\exp(-k)\left[1 + k + \frac{k^2}{2!} + \frac{k^3}{3!} + \frac{k^4}{4!} + \frac{k^5}{5!} + \cdots + \frac{k^n}{n!} \right] = 1.0$$

Such a property means that we have obtained a probability mass function. Each term within the brackets must be a probability value.

It has been found that this distribution approximates very well to the random occurrence of events, measured over relatively long periods of time, when the value for k is taken to be the expected, or average, number of occurrences.

This expression is known as the Poisson distribution and is usually written in the more general form:

$$p_X(x) = \frac{\exp(-k) \cdot k^x}{x!}$$

Example 3.8

Observations of a river that is subject to flooding have been carried out over a 100 year period. The number of floods that have occurred per year are given in the following table.

Number of years	Number of floods
30	0
35	1
22	2
10	3
3	4
0	5

Determine the probability of flooding for the next fifty years.

Solution

Total number of floods $= 35 + 2 \times 22 + 3 \times 10 + 3 \times 4 = 121$
Therefore, expected value of $k = 121/100 = 1.21$
Now, $\exp(-1.21) = 0.2982$

Writing out the full equation to six terms gives:

$$= 0.2982\left(1 + 1.21 + \frac{1.21^2}{2!} + \frac{1.21^3}{3!} + \frac{1.21^4}{4!}\right)$$

$$= 0.2982 + 0.3608 + 0.2183 + 0.0881 + 0.0266$$

These terms, when taken in successive order, represent the probabilities of no floods, one flood, two, three and four floods.

Each probability value can be found separately by using the general formula by denoting X as the random discrete variable that represents the number of floods possible. For example:

$$\text{Probability of 3 floods} = p_X(3) = \frac{\exp(-1.21)1.21^3}{3!}$$

$$= 0.0881$$

Using the probability values obtained the expected flood situation for the next fifty years can be estimated.

$$
\begin{aligned}
\text{No. of years with no floods} \quad &= 0.2982 \times 50 = 14.9 \\
\text{No. of years with one flood} \quad &= 0.3608 \times 50 = 18.1 \\
\text{No. of years with two floods} \quad &= 0.2183 \times 50 = 10.9 \\
\text{No. of years with three floods} &= 0.0881 \times 50 = 4.4 \\
\text{No. of years with four floods} \quad &= 0.0266 \times 50 = 1.3
\end{aligned}
$$

The reader may observe that the calculated years only summate to 49.6. This is because the $p_X(x)$ values summate to 0.992 instead of 1.0, inferring that there is a minimal probability (0.008) of five floods occurring.

Alternative solution – by the use of tabulated values. Values of:

$$F_X(x) = \sum_{r=0}^{x} \frac{\exp(-k)k^x}{k!}$$

are tabulated for k up to 40 and x up to 60. (See Neave, 1978.)
    The portion of tabulation relevant to the example is set out below.

k	x	0	1	2	3	4	5	6	7	8
1.2		0.3012	0.6626	0.8795	0.9662	0.9985	0.9997	1.0000		
1.3		0.2725	0.6268	0.8571	0.9569	0.9893	0.9978	0.9996	0.9999	1.0000

By linear interpolation between the two sets of values it is possible to determine $F_X(x)$ values for $k = 1.21$:

x	0	1	2	3	4	5	6	7
$F_X(x)$	0.2983	0.6590	0.8773	0.9653	0.9920	0.9984	0.9997	1.0000

Hence:

$$
\begin{aligned}
P[X = 0] &= 0.2983 & p_X(0) &= 0.2983 \\
P[X \leqslant 1] &= 0.6590 & p_X(1) &= 0.3607 \\
P[X \leqslant 2] &= 0.8773 & p_X(2) &= 0.2183 \\
P[X \leqslant 3] &= 0.9653 & p_X(3) &= 0.0880 \\
P[X \leqslant 4] &= 0.9920 & p_X(4) &= 0.0267 \\
P[X \leqslant 5] &= 0.9984 & p_X(5) &= 0.0064 \\
P[X \leqslant 6] &= 0.9997 & p_X(6) &= 0.0013
\end{aligned}
$$

It is seen that calculated values of $p_X(x)$ compare very well with those derived from the tabulated values of $F_X(x)$.

THE MEAN AND VARIANCE OF A POISSON DISTRIBUTED VARIABLE

It can be shown that if X is a Poisson distributed discrete variable then $m_X = \text{Var}(X) = k$.

Example 3.9

The number of deaths caused by accidents on a stretch of motorway have been measured for the 30 years since its construction. The results are shown in the table below.

Deaths	Years
0	15
1	10
2	4
3	1

If X is the random variable representing the number of deaths, show that it has a Poisson distribution and determine its mean value and standard deviation.

Solution
The example will be solved numerically.

$$k = \tfrac{21}{30} = 0.7; \qquad \exp(-0.7) = 0.4965$$

Assuming a Poisson distribution the following table can be obtained:

x	$p_X(x)$	Frequency $= 30p_X(x)$	$xp_X(x)$	$x^2p_X(x)$
0	0.4965	14.9	0.0	0.0
1	0.3476	10.4	0.3476	0.3476
2	0.1217	3.7	0.2434	0.4868
3	0.0284	0.9	0.0852	0.2556
4	0.0050	0.2	0.0200	0.0800
5	0.0007	0.0	0.0035	0.0175
Total	1.0000		0.6997	1.1875

It is seen that the calculated frequencies agree very well with those actually observed. The assumption that X is a Poisson distributed variable is therefore satisfactory. From the table:

$$E[X] = m_X = 0.6997$$

and:

$$\text{Var}(X) = 1.1875 - 0.6997^2 = 0.6979$$

Both these values are extremely close to 0.7 which is the value of k which proves, numerically, that m_X equals $\text{Var}(X)$ and is equal to k.

THE POISSON APPROXIMATION TO THE BINOMIAL DISTRIBUTION

The binomial distribution can closely approximate to the Poisson distribution when it consists of rare events, i.e. when n is large and p is small.

An event is generally considered rare when n is greater than or equals 50 and np is less than five. For such instances the binomial probability distribution can be written as:

$$p_X(x) = \frac{\exp(-k)k^x}{x!}$$

where $k = m_X = np$.

The normal distribution

The probability density functions of continuous random variables can have various forms, as illustrated in chapter 2, but the most common, particularly in engineering, is the humped back, or single peaked, distribution with the maximum number of values tending to occur towards the centre of the curve.

This bell-shaped distribution curve is known as the normal distribution curve and has the equation:

$$f_X(x) = \frac{1}{\sigma_X\sqrt{2\pi}} \exp[-(x - m_X)^2/2\sigma_X{}^2] \qquad -\infty \leqslant x \leqslant \infty$$

The mean and the variance are considered to be the parameters of the normal distribution. Hence a normally distributed random variable, X, would be described as:

$$X = N(m, \sigma^2)$$

Example 3.10

The undrained shear strength of a clay has a mean value of 44.85 kN/m^2 and a standard deviation of 6.56 kN/m^2. Fit these values to a normal distribution.

Solution

Substituting for m_X and σ_X in the formula gives:

$$f_X(x) = \frac{1}{16.44} \exp[-(x - 44.85)^2/86.07]$$

By inserting values for x into the above formula the corresponding values of $f_X(x)$ can be obtained. For example:

$$\text{for } x = 30 \qquad f_X(x) = 0.0047$$

If values for x, covering an appropriate range, are inserted into the equation, the normal distribution is obtained.

Example 3.11

Using the results of example 3.10, determine the probability that the value of X will lie within the range 40 to 50 kN/m^2.

Solution
Values of x and the corresponding values of $f_X(x)$ over the range $x = 40$ to $x = 50$ kN/m^2 are out below.

x	40	41	42	43	44	45	46	47	48	49	50
$f_X(x)$	0.0463	0.0512	0.0553	0.0584	0.0603	0.0608	0.0599	0.0576	0.0542	0.0498	0.0447

By Simpson's rule, the area under the curve is 0.554, i.e. probability = 0.554.

(The application of Simpson's rule is described in appendix I.)

Standardised variables

The properties of the normal distribution are well documented and tables of values of $f_X(x)$, etc. are readily available. In order to use this information it is first necessary to change the x values into a set of standardised, or reduced, variables. These variables are dimensionless and are usually given the letters 'z' or 'y'. In reliability analysis work the letter 'z' is used to designate the limit state function and it is therefore probably less confusing if the letter 'y' is used for reduced variables.

The corresponding reduced variable, y, of a random variable, X, which has a standard deviation, σ_X, and has attained a value, x, is given by the expression:

$$y = \frac{x - m_X}{\sigma_X}$$

The pdf of y is known as the standardised normal density function and has the property that its mean is zero and its variance is one. Because of this, the pdf is often symbolised as $N(0, 1)$ whereas the pdf of the normal distribution of example 3.9 is written as $N(44.85, 6.56^2)$.

$$f_Y(y) = \frac{1}{\sqrt{2\pi}} \exp(-y^2/2)$$

(See Fig. 3.2.) Values of both $f_Y(y)$ and of $F_Y(y)$ (the cumulative density function, symbol cdf), are listed in appendices II and III respectively.

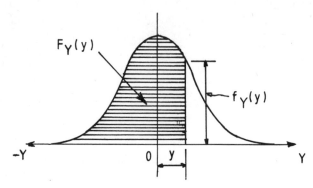

Fig. 3.2 The standardised variable Y

Example 3.11 will now be solved using standardised variables.

$$y_{40} = (40\text{–}44.85)/6.56 = -0.7393$$

Now:

$$P[Y = 40] = F_Y(40)$$

and, from appendix III:

$F_Y(40) = 0.230$

Also:

$$y_{50} = (50\text{–}44.85)/6.56 = 0.7851$$

$$P[Y = 50] = F_Y(50)$$

and, from appendix III:

$$F_Y(50) = 0.784$$

Hence the probability that X will lie between 40 and 50 kN/m^2 is:

$$0.784 - 0.230 = 0.554$$

THE NORMAL APPROXIMATION TO THE BINOMIAL DISTRIBUTION

If n is small, values for a binomial distribution are calculated directly or obtained from tables. This is difficult if n is greater than 100.

If X is a binomially distributed discrete random with p and q values which are both not too close to zero, then its distribution becomes very close to a normal distribution if n is increased. The mean of this distribution is np and its standard deviation is $\sqrt{npq}$.

This approximation to the normal distribution is exact when n is infinite and is very close when both np and nq are greater than five. Therefore, we can use the tabulated properties of the standardised normal distribution to determine approximate values for a binomial distribution. In this case the standardised variable, y, becomes:

$$y = \frac{X - np}{\sqrt{npq}}$$

When a curve tends to approach another line at infinity the curve is said to be asymptotic to the line. This is why the binomial distribution is often referred to as being asymptotically normal.

There is one problem that arises with approximating a continuous to a discrete variable. The discrete variable can only achieve a value taken from a finite set of values, such as $x_1, x_2, x_3, x_4, \ldots, x_n$, whereas the continuous variable that is to approximate it can have any value from an infinite set ranging from x_1 to x_n.

As discussed in chapter 2 the expression $P[X = x]$ has a meaning when the variable is discrete but is equal to zero if the variable is continuous. Hence, in order to approximate a discrete variable to a continuous one, we must think of each discrete value as being the centre point of the corresponding histogram cell (Fig. 3.3).

Considering the binomial distribution – if the number of Bernoulli trials was 200 and if X was the number of successes, then the probability that there were 125 successes, $P[X = 125]$, must be approximated to $P[124.5 \leqslant x \leqslant 125.5]$.

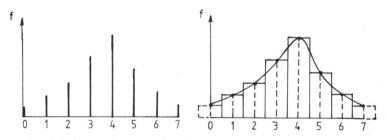

Fig. 3.3 Approximation of a discrete to a continuous distribution

Example 3.12

X is a discrete random variable with the distribution $B(45, 0.35)$. Find the following probabilities: (i) $P[X = 20]$; (ii) $P[22 \leqslant X \leqslant 25]$ by (a) formulae; (b) using a normal approximation.

(Note that tabulated values cannot be used as n is greater than 20.)

Solution

(ia) – using formulae

$$p = 0.35 \Rightarrow q = 0.65$$

$$P[X = 20] = \binom{45}{20} 0.35^{20} \times 0.65^{25} = \frac{45!}{20!\,25!} 0.35^{20} \times 0.65^{25}$$

$$= 0.0507$$

(ib) – by normal approximation

$$np = 0.35 \times 45 = 15.75 \qquad \sqrt{npq} = \sqrt{0.35 \times 0.65 \times 45} = 3.200$$

$$P[X = 20] \doteqdot P[19.5 \leqslant X \leqslant 20.5]$$

$$y_{19.5} = \frac{19.5 - 15.75}{3.200} = 1.172$$

$$y_{20.5} = \frac{20.5 - 15.75}{3.200} = 1.484$$

From appendix III, $F_Y(1.172)$ equals 0.8794 and $F_Y(1.484)$ equals 0.9310. Hence:

$$P[X = 20] \doteqdot 0.9310 - 0.8794 = 0.0516$$

(iia) – using formulae

$$P[22 \leqslant X \leqslant 25] = P[X = 22] + P[X = 23]$$

$$+ P[X = 24] + P[X = 25]$$

$$= \binom{45}{22} 0.35^{22} 0.65^{23} + \binom{45}{23} 0.35^{23} 0.65^{22}$$

$$+ \binom{45}{24} 0.35^{24} 0.65^{21} + \binom{45}{25} 0.35^{25} 0.65^{20}$$

$$= 0.01910 + 0.01029 + 0.00508 + 0.00230 = 0.037$$

(iib) – by normal approximation

$$P[22 \leqslant X \leqslant 25] \doteqdot P[21.5 \leqslant X \leqslant 25.5]$$

Now, $y_{21.5} = 1.80$ and $y_{25.5} = 3.047$ and, from appendix III, we can obtain values for the two cumulative probabilities:

$$F_Y(1.80) = 0.9641 \qquad F_Y(3.047) = 0.9989$$

Hence:

$$P[22 \leqslant X \leqslant 25] \doteq 0.9989 - 9.9641 = 0.035$$

THE NORMAL APPROXIMATION TO THE POISSON DISTRIBUTION

The Poisson distribution can be approximated to the normal distribution with a procedure identical to that just described for the binomial distribution except that the standardised variable, y, is found from the expression:

$$y = \frac{X - k}{\sqrt{k}}$$

Example 3.13

X is a discrete random variable with a mean value of 12 and a Poisson distribution. Determine: (i) $P[X = 10]$; (ii) $P[X \leqslant 10]$; (iii) $P[10 \leqslant X \leqslant 14]$.

Solution
For illustrative purposes each probability value will be determined in three different ways: (a) by formulae; (b) from tabulated-cdf values; and, (c) by normal approximation.

(ia) $$P[X = 10] = p_X(10) = \frac{\exp(-12)12^{10}}{10!} = 0.1048$$

(ib) $$P[X = 10] = F_X(10 - F_X(9)$$

From Neave's (1978) or similar tables using $k = 12$:

$$= 0.3472 - 0.2424$$

$$= 0.1048$$

(ic) $P[X = 10]$ must be approximated to $P[9.5 \leqslant X \leqslant 10.5]$. Now:

$$y_{10.5} = \frac{10.5 - 12}{\sqrt{12}} = -0.4330$$

$$y_{9.5} = \frac{9.5 - 12}{\sqrt{12}} = -0.7217$$

Hence, from appendix III:

$$P[X = 10] \doteq 0.3325 - 0.2352$$

$$= 0.0973$$

(iia) $P[X \leqslant 10] = p_X(0) + p_X(1) + p_X(2) + \cdots + p_X(10$

$$= \frac{\exp(-12)12^0}{0!} + \frac{\exp(-12)12^1}{1!}$$

$$+ \frac{\exp(-12)12^2}{12!} + \cdots + \frac{\exp(-12)12^{10}}{10!}$$

$$= 0.3472$$

(iib) $P[X \leqslant 10] = F_X(10) = 0.3472$ from tables

(iic) $P[X \leqslant 10] \doteqdot P[X = 10.5]$

Now:

$$y_{10.5} = \frac{10.5 - 12}{\sqrt{12}} = -0.4330$$

Hence, from appendix III:

$$P[X \leqslant 10] \doteqdot 0.3325$$

(iiia)

$P[10 \leqslant X \leqslant 14] = p_X(10 + p_X(11) + p_X(12) + p_X(13) + p_X(14)$

$$= \frac{\exp(-12)^{10}}{10!} + \frac{\exp(-12)^{11}}{11!} + \cdots + \frac{\exp(-12)^{14}}{14!}$$

$$= 0.5297$$

(iiib) $P[10 \leqslant X \leqslant 14] = F_X(14) - F_X(9)$

From tables:

$$= 0.7720 - 0.2424$$

$$= 0.5296$$

Note that, as X is a discrete variable, $F_X(9)$ and not $F_X(10)$ is used.

(iiic) $P[10 \leqslant X \leqslant 14] \doteqdot P[9.5 \leqslant X \leqslant 14.5]$

$$y_{14.5} = \frac{14.15 - 12}{\sqrt{12}} = 0.7271$$

$$y_{9.5} = \frac{9.5 - 12}{\sqrt{12}} = -0.7217$$

Hence, from appendix III:

$$P[10 \leqslant X \leqslant 14] \doteqdot 0.7648 - 0.2352$$
$$= 0.5296$$

The central limit theorem

One of the most important theorems of statistics is the central limit theorem and it is extremely important in sampling theory.

If X is a population with a mean value m_X and a standard deviation σ_X and if $S_n = x_1, x_2, x_3, \ldots, x_n$ is a random sample collection from it, then the central limit theorem states that:

The mean of the sample, m_S, given by $\sum_{i=1}^{i=n} x_i$, is really a variable with a distribution that approximates to normal and has a mean value m_X and a variance σ_X^2/n.

In other words:

$$m_S = N(mX, \sigma_X^2/n)$$

The approximation improves as the value of n is increased.

In practical terms, even for only a small number of variables, provided that they are more or less of the same weight and that any dependency is of a small degree, the distribution of their sum will tend to be normal.

Sampling theory

For the test results obtained from a set of samples to be meaningful, two main conditions must be satisfied:

(i) The samples must be representative of the material from which they were taken.

   If the sample is a housewife interviewed about her political views as part of a survey of local opinion, then she must be representative of other housewives in the district. Similarly if a sample of sand has been selected from a lorry's delivery for testing, then the sample should be representative of the main body of the sand.

(ii) The number of samples must be sufficient.

Point estimation

In structural engineering it is generally not too difficult to obtain an adequate number of measurements for the strengths and unit weights of the constructional materials involved so that reasonable estimates of the means and standard deviations of these parameters are possible.

However, in soil engineering, at the very best, it is usual for only one statistical sample of some four, five or six soil samples to be collected for each type of soil encountered during a site investigation. This means that the value of the mean of a population, m_X, has to be estimated from only one sample value, m_S. Such an estimation is called a point estimation.

In such cases it is often useful to be able to estimate how far away the sample value is from the actual mean value of the population. This is achieved by a term called the standard error of the mean.

THE STANDARD ERROR OF THE MEAN

As discussed, the value of m_S in soils is usually determined from a small group of measured values.

Let us assume that there is a large number of measurements and that these values are grouped into sets of five by some form of random selection. If m_S is determined for each group then several different values of m_S will be determined. The distribution of these m_S values will be found to be distributed about a mean value, m_X, and to have a certain variance value. For all practical purposes m_X is the value of the mean of the population.

If the values had been placed in groups of ten, the resulting m_S values would still have been distributed about the same mean value, m_X, but would have had a smaller variance (see Fig. 3.4). In other words the variances of equal groups of values is related to n, the number of values in each group and we can say:

$$\text{Var}_n = \frac{\sigma^2}{n}$$

where:

Var_n = variance of values when grouped with n values per group
σ^2 = variance of values when each value is taken separately

Now the standard deviation is the square root of the variance. Thus:

$$\sigma_n = \frac{\sigma}{\sqrt{n}}$$

The term $\sigma/\sqrt{n}$ is the standard error of the mean.

From the central limit theorem we know that if a random sample of n values is selected from a population that has a mean value m_X and a variance of σ^2 then, as n increases, the distribution of:

$$\frac{S_n - nm_X}{\sigma\sqrt{n}} = \frac{S_n/n - m_X}{\sigma\sqrt{n}/n} = \frac{m_S - m_X}{\sigma/\sqrt{n}}$$

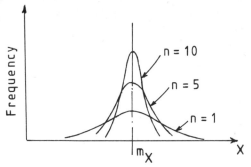

Fig. 3.4 Frequency curves of mean values of groups

tends to be normally distributed. In other words the sampling distribution of the means approaches a normal distribution as the value of n increases.

Therefore, the properties of the normal curve can be used to estimate the dependability of a point estimation of the mean of a population. The smaller the value of $\sigma/\sqrt{n}$, i.e. the larger the value of n, the more reliable the estimated value of the population mean.

Now, for a specified level of confidence, m_X must lie within the range of values $m_S \pm k(\sigma/\sqrt{n})$ where k is a constant dependent on the level of confidence. (The level of confidence is equal to the unhatched area of Fig. 3.5.)

If, for example the required probability was 75% then the areas of the truncated ends of the distribution (shown hatched in Fig. 3.5) are both equal to $0.5(100 - 75) = 12.5\%$.

From appendix III it is seen that the y values that give values for $F_Y(y)$ of 12.5 and 87.5 are -1.15 and 1.15 respectively.

There is a 75% probability that m_X will lie within the range

$$m_S \pm 1.15\sigma/\sqrt{n}$$

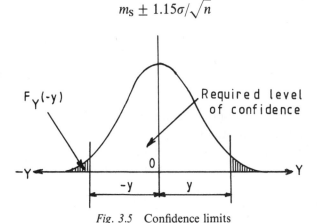

Fig. 3.5 Confidence limits

Example 3.14

A sample of values taken from a normally distributed population are 6.8, 7.2, 5.4, 8.8 and 10.2.

    If the standard deviation of the population is 2.1 determine the range of the sample values within which the mean of the population has a 95% probability of lying.

Solution

For a level of confidence of 95%, $F_Y(y)$ equals 2.5 and 97.5%. Hence, y equals ± 1.96 (from appendix III) and m_S equals 7.7. Therefore, the range of sample values within which m_X has a 95% chance of occurring is given by:

$$7.7 \pm 1.96 \times \frac{2.1}{\sqrt{5}} = 5.86 \text{ to } 9.55$$

Note: Usually the standard deviation of a population must be estimated from the test results available using the Bessel correction described in chapter 2.

Student's *t* distribution

Let the ratio:

$$\frac{\text{error in the sample mean}}{\text{standard error of the mean}} = y$$

Then:

$$y = \frac{|m_S - m_X|}{\sigma/\sqrt{n}} = \frac{|m_S - m_X|}{\sigma}\sqrt{n}$$

where y is the standardised error between the sample mean and the population mean.

    Generally an estimated value, $\hat{\sigma}$, has to be used in place of σ which means that, instead of establishing a variable Y we establish a variable T where:

$$t = \frac{|m_S - m_X|}{\hat{\sigma}/\sqrt{n}}$$

or:

$$t = \frac{|m_S - m_X|\sqrt{n-1}}{s}$$

where s is the standard deviation of the n sample values.

T has a distribution similar to, but not the same as, the normal distribution and was discovered by W. S. Gosset who published his findings under the pseudonym A. Student.

The value of $\hat{\sigma}$ gets closer to the value of σ as the value of n is increased. For the value of n equals 30, the t distribution becomes identical to the normal distribution.

In geotechnical work there is little chance of the value of n approaching anything like 30 and it is therefore more realistic to accept the fact that the probability values of the t distribution should be used in place of values obtained from the normal distribution for geotechnical work.

There is little need for the reader to become involved with the mathematics that generate the t distribution as appendix IV tabulates relevant $F_T(t)$ values. With the use of this table the t value corresponding to a particular probability (the $F_T(t)$ value) can be found provided that the degrees of freedom value, v, is also known.

The degrees of freedom of a set of values is the number of those values that can be of any magnitude, within the constraint of the calculations about to be carried out on them. For this particular aspect of statistics we can say that v equals $(n - 1)$.

Example 3.15

(i) A sample consists of the following values:

$$10, \quad 8, \quad 6, \quad 4$$

Determine the range of values within which the true value of the mean of the population has a 97.5% probability of lying.

(ii) In order to increase accuracy a further four values are obtained. These are:

$$4.6, \quad 6.8, \quad 8, \quad 9.5$$

With this extra information determine the range of values within which the mean value has a 97.5% probability of lying.

Solution

The sample sizes for both (i) and (ii) are small (n being less than 30) and we must therefore use the properties of the t distribution.

Example 3.13 illustrates that, for a normal distribution, the two end probabilities would be 1.25% and 98.75% (giving 97.5% between them). These $F_Y(y)$ values correspond to $y = \pm 2.24$.

With the t distribution the value of t is always positive so we must

therefore determine the value of t that corresponds to a probability value
of 97.5%.

(i) $$m_S = \frac{10 + 8 + 6 + 4}{4} = 7.0; \qquad s = 2.236$$

Thus:

$$\hat{\sigma} = \sqrt{\tfrac{4}{3}} \times 2.236 = 2.58$$

From appendix IV, for $v = 3$ and $P = 0.975$, $t = 3.18$. Therefore the
range of values within which the population mean will lie is:

$$7.0 \pm 3.18 \times \frac{2.58}{\sqrt{4}} = 2.90 \text{ to } 11.10$$

(ii) $$m_S = \frac{4.0 + 4.6 + 6.0 + 6.8 + 8.0 + 8.0 + 9.5 + 10}{8} = 7.11;$$

$$\hat{\sigma} = 2.170$$

From appendix IV, for $v = 7$ and $P = 0.975$, $t = 2.37$. Therefore the
range of values within which the population mean will lie (for $P = 95\%$)
is:

$$7.11 \pm \frac{2.37 \times 2.170}{8} = 5.29 \text{ to } 8.93$$

Minimum sample number

From the soil samples obtained during the first phase of a site
investigation it is possible to determine whether or not the number of
samples collected from each sub-region was sufficient for the required
accuracy of prediction. If not then the second phase of the site
investigation will be necessary so that further samples can be obtained.

The minimum number of soil samples that will have to be collected
from a sub-region depends on various factors, not least being the
accuracy of prediction asked for by the design engineer. If he demands
that the average value of the test results should equal the average in-situ
value he is demanding the impossible as only an infinite number of
samples could satisfy this condition.

As illustrated in the preceding example, with the arithmetic mean and
standard deviation of a single statistical sample, it is possible to evaluate
the range of values in which, within a certain probability of say 95% or
99%, the mean value of the population will lie. The number of values that
will have to be obtained, for the one statistical sample, will depend on the
acceptable width of this range.

Example 3.16

Four undisturbed samples were taken from a stiff clay deposit. Undrained triaxial tests on these samples gave the following values for c_u, the undrained shear strength values:

$$102, \quad 98, \quad 95, \quad 109 \quad (kN/m^2)$$

Determine the minimum number of samples of the clay that should be taken so that, within a 95% probability, the average in-situ undrained shear strength value will be within 5% of the mean test result.

Solution

With the four samples v equals 3. From appendix IV, for P equal to 0.95, t is 2.35. Now the mean c_u test result was $101\ kN/m^2$ and $\hat{\sigma}$ was $6.06\ kN/m^2$. Therefore, the range of values for P equal to 95% is given by:

$$101 \pm 2.35 \times 6.06/\sqrt{4} = 93.9 \text{ to } 108.1\ kN/m^2$$

(This is some 7% either side of the mean.) Obviously more samples are required if the range is to be only 5% on either side of the mean, i.e. from 96.0 to 106.1.

Try five samples:
Then v equals 4 and from appendix IV for P equal to 95%, t is 2.13. Therefore, the range of values from the mean is:

$$\frac{2.13 \times 6.06}{5} = 5.77\ kN/m^2 \quad (= 5.7\%)$$

Try six samples:
Then v equals 5 and, from appendix IV, t is 2.02. Therefore, the range of values from the mean is:

$$\frac{2.02 \times 6.06}{6} = 5.00\ kN/m^2 \quad (= 5\%)$$

At least two further samples of the clay must be obtained and tested.

It should be noted that to estimate the variance of the population to the same precision as the mean requires a much larger sample and is not usually attempted for soil and rock reliability analyses.

Characteristic values

There are two different types of confidence limit and it is important that the reader appreciates this.

Examples 3.14, 3.15 and 3.16 deal with the problem of estimating the range of values within which the true mean value of a population is likely to lie, to a specified level of confidence, i.e. to a specified value of probability. Significance tests such as used in these examples, when both ends of the range of possible values are considered, are known as two-sided tests.

In civil engineering we are often concerned with the most probable maximum loadings, or the most probable minimum strengths. We may consider the most probable minimum strength to be the value that has only a 5% probability of occurrence. The level of confidence is still 95% but as we are only considering the bottom end of the distribution, the y value we use will be that corresponding to an $F_Y(y)$ value of 5%: we are using a one-sided significance test.

Alternatively the most probable maximum loading may be considered to have a 5% chance of occurrence and we would consider this loading to occur at a z value corresponding to $F_Y(y)$ equals 95%, again a one-sided significance test.

This approach has led to the use of characteristic values, suggested in CP110 (1972), for reinforced concrete design. This code defines the characteristic strength as the value of the cube strength of the concrete, the yield or proof stress of the reinforcement, or the ultimate load of a prestressing tendon, below which not more than 5% of the test results will fall.

For a parameter with a normal distribution:

Characteristic strength = mean value − 1.645 × standard deviation

Characteristic load   = mean value + 1.645 × standard deviation

The draft standard on the reliability of structures, produced in 1984 by the International Standards Organization (ISO) suggests that characteristic values could be used in soil calculations.

Exercises

PERMUTATIONS AND COMBINATIONS

3.1 Evaluate:

(i) $_{15}P_4$; (ii) $_{15}P_1$; (iii) $_8P_3$; (iv) $_6C_3$; (v) $_6C_6$; (vi) $_{28}C_{20}$.

Answer: (i) 32 760; (ii) 15; (iii) 336; (iv) 20; (v) 1; (vi) 3 108 105

3.2 Five soil samples are to be tested, one after the other. In how many different ways can the samples be selected for test?

Answer: 120.

3.3 Twelve concrete cubes have been delivered to a materials laboratory where any six of them are to be tested. How many arrangements of six different test cubes are possible?

Answer: 924.

3.4 Determine the number of different arrangements that can be made from the letters that occur in the word 'STATISTICS'.

Answer: 50 400.

3.5 Six seats are placed around a table.
(i) In how many ways can six people sit around this table so that at least one is in a different seat in each arrangement?
(ii) If the actual seat positions are ignored how many different arrangements of the six people around the table are possible?

(For part (ii) it should be remembered that if the actual seat positions are ignored then the first person may sit anywhere. The second person has therefore the choice of five seats, the third person the choice of four seats, etc. Hence, the total number of possible arrangements is 5!)

Answer: (i) 720; (ii) 120.

3.6 Thirty-two samples are to be divided into groups of eight. In how many ways can this division be made?

Answer: 9.956×10^{16}.

BINOMIAL COEFFICIENTS

3.7   Determine the coefficients of the third and seventh terms in the expansion of $(x + y)^{10}$.

Answer: 45 and 210.

3.8   Determine the binomial expansion for $(x + 2)^4$ and hence, by putting $X = 2$, show that $4^4 = 256$.

BINOMIAL DISTRIBUTION

3.9   A fair die is thrown eight times. Determine the probability of:
(i) Exactly three 2s being obtained in the eight throws.
(ii) At least three 2s will be obtained.

Answer: (i) 0.104; (ii) 0.135.

3.10   If X is a discrete random variable with the probability distribution B(50, 0.4) determine the probability that X will equal 25, (i) by formulae, and (ii) by approximation to the normal distribution.

Answer: (i) 0.0405; (ii) 0.0408 ($z_{24.5} = 1.2991$ and $z_{25.5} = 1.5878$).

Chapter Four

The Second Moment Method of Reliability Analysis

The probability of failure of a structure

The term 'failure' is used here in its most general sense and implies the failure of the structure to satisfy some particular limit state criterion, which may or may not be actual structural failure.

The frequenistic approach cannot be applied to the estimation of the probability of failure of a civil engineering structure where the design and construction is a once only operation. Even for similar, or prefabricated structures, where aspects of the design work may be repeated, each structure will be built on a different site leading to the possibility of different soil and geological conditions.

An evaluation of the probability of failure of a structure must, therefore, be undertaken by the application of statistics and probability theory.

Consider the resistance or strength of a structure, R, and the applied

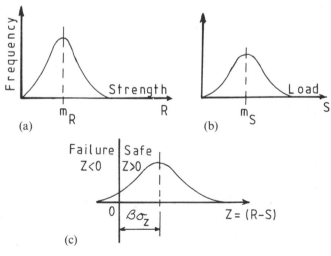

Fig. 4.1   The reliability index β

loading, S, to which it will be subjected. The values of both R and S are not fixed but will assume any value within a range of values. The extent of these ranges will vary with the dgree of probability decided as acceptable for the design problem (usually 95%). R and S are, therefore, random variables with definitive, although possibly unknown, probability density functions (pdfs).

Figures 4.1a and b show assumed pdfs for R and S and illustrate that failure will occur when R is less than S. If Fig. 4.1b is subtracted from Fig. 4.1a then the probability curve of $Z = R - S$ (strength minus load), is obtained, Fig. 4.1c.

The probability of failure, P_f, equals $P[(R - S) \leqslant 0]$ which is equal to $P[Z \leqslant 0]$ and Z is the limit state function for the particular mode of failure being considered.

Methods of reliability analysis

There are three main methods in which a structure may be designed to achieve a certain probability of failure value and these are described in Report 63 of the C.I.R.I.A. (1976):

Level I   a design method involving characteristic values and partial factors of safety.

Level II   a reliability analysis which uses safety checks at a selected point (or points) on the failure boundary, defined by Z, the appropriate limit state function.

Level III an extremely comprehensive probability analysis in which 'exact' safety checks, using a full distributional approach for the variables, are carried out for the whole structural system.

The level I approach is more a hope for the future than a method that exists at the moment. When evolved, the method will not require the actual determination of a value for P_f as a particular limit state will be considered safe if the appropriate partial safety factors are not exceeded. A list of these factors will be given in the design codes.

The problem is that values for these partial safety factors, for a suitable range of structural elements, will first have to be obtained from reliability analyses carried out by either a level II or a level III approach.

The level III method is really a form of pure mathematics and will probably only be used for the analysis of special structures, in which the reliability level is of critical importance or where it is particularly important to optimise the design.

The level II method, often referred to as the reliability index method, involves fairly straightforward mathematics and there is general agreement that this method has the potential of either being used for the evaluation of suitable partial factors for the level I method or as a direct design method in its own right.

This book deals only with the level II method.

Reliability index

Generally, there is not sufficient information regarding the tails of the Z distribution and the criterion P_f equals $P[Z \leqslant 0]$ is therefore replaced with one that involves the mean value and standard deviation of Z.

In Fig. 4.1c the distance from the mean of Z, m_z, to the failure boundary, i.e. the point at which Z equals 0, can be expressed in terms of σ_z, the standard deviation of Z, and equals $\beta \sigma_z$. β is known as the reliability index and is a measure of the safety of the system. Obviously:

$$m_Z - \beta \sigma_z = 0 \qquad \text{i.e. } \beta = m_z / \sigma_z$$

Now:

$$m_Z = m_R - m_S$$

Hence:

$$\beta = \frac{m_R - m_S}{\sigma_Z}$$

The factor of safety, F, is equal to m_R / m_S.

The expression for F is purely deterministic whereas the expression for β includes not only m_R and m_S but also σ_Z, a measure of the uncertainty of both R and S. It can therefore be seen that β is a more meaningful measure of safety than F.

Basic variable space

In most practical problems R and S will rarely be single variables and will be vectors made up from the set of relevant basic variables.

Basic variables are the fundamental parameters involved in the design. Examples are the ultimate strengths of the materials to be used, the intensity and type of loadings, depth of reinforcement, etc. If n basic variables make up a particular random variable X such that:

$$X = (X_1, X_2, X_3, \ldots, X_n)$$

then the basic variable space is the n-dimensional space that will represent all possible values of X.

This means that x, which equals $(x_1, x_2, x_3, \ldots, x_n)$ is a single point of coordinates $x_1, x_2, x_3, \ldots, x_n$ and represents the situation when the basic variables $X_1, X_2, X_3, \ldots, X_n$ have values x_1 to x_n.

Z is a function of all the relevant basic variables so we can say that, generally:

$$Z = g(X_1, X_2, X_3, \ldots, X_n)$$

and that:

$$P_f = P[Z \leqslant 0] = P[g(X_1, X_2, X_3, \ldots, X_n) \leqslant 0]$$

Example 4.1

A granular soil will be subjected to a shear stress, τ. The normal stress on the shear plane, σ, will have a mean value of $100 \, kN/m^2$ and a standard deviation of $20 \, kN/m^2$.

The angle of friction of the soil has a mean value of $35°$ and a standard deviation of $5°$.

Plot the failure boundary in the basic variable space and determine the reliability index of the system if τ has a fixed value of $50 \, kN/m^2$.

Solution

Coulomb's law of soil shear strength states that, for a granular soil:

$$\tau = \sigma \tan \phi$$

Therefore:

$$Z = \sigma \tan \phi - \tau = \sigma \tan \phi - 50$$

Z can be represented as $g(\sigma, \phi, \tau)$ or $g(X_1, X_2, C)$ where X_1 equals σ, X_2 equals ϕ and C is a constant (equal to 50).

As there are only two random variables, n is 2. The failure boundary will show up as a line on a two dimensional plot. This line can be obtained from the equation $\tan \phi$ equals $50/\sigma$ which is purely deterministic and has nothing to do with probability theory. However, if the scales for σ and ϕ are so chosen that the length representing one standard deviation of σ $(20 \, kN/m^2)$ is equal to the length that represents one standard deviation of ϕ $(5°)$ then the minimum distance from the mean point to the failure boundary will be equal to β, the reliability index.

By selecting suitable values for σ a range of corresponding $\tan \phi$ values can be obtained which leads to the values of ϕ tabulated below.

kN/m²	40	60	80	100	120	140	160
tan ϕ	1.25	0.833	0.625	0.5	0.417	0.357	0.313
ϕ (degrees)	51.3	39.8	32.0	26.6	22.6	19.6	17.4

The failure boundary is shown in Fig. 4.2a together with the mean point
(100, 35°). The central part of the diagram is shown enlarged in Fig. 4.2b
so that an accurate determination of the distance from the mean point to
the failure boundary can be obtained. Due to the scales chosen, this
distance is in terms of the standard deviation of Z and is found to be
equal to 1.16. Hence the reliability index of the system is 1.16.

It can be seen that the smaller the minimum distance, the nearer the
mean point is to the failure boundary and the greater the risk of failure.

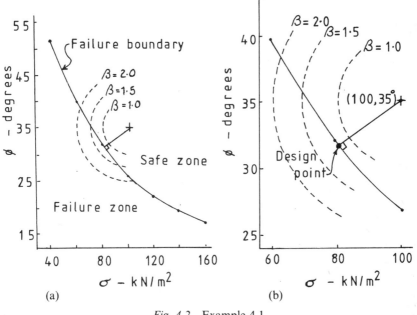

Fig. 4.2   Example 4.1

Note: There is an unfortunate clash of symbols in the example. Civil
engineers use the symbol σ to represent normal stress whereas statis-
ticians use the same symbol to represent standard deviation. As there will
generally be little risk of confusion it was decided to leave the symbol σ to
represent both normal stress and standard deviation in this text.

Reliability analysis by the second moment method

When there are several variables the failure boundary is a surface, not a
line, and the plotting technique just described cannot be used.

A practical alternative is the level II method which, as it deals with means and variances, is classified as a second moment approach. (The variance of a random variable is its second central moment.)

Second moment methods of reliability analysis originate from work by Mayer (1926), but were not seriously considered for a further forty years when the works of Cornell (1969), Ravindra et al. (1969) and Rosenblueth and Esteva (1972) were published.

A simple method of safety checking, which involved some statistical measure of the uncertainties but did not employ complex integrations using full probabilitity distributions was suggested. The mathematical work was considerably simplified by assuming a planar failure boundary. A linear failure boundary rarely occurs in practice (see, for example, Fig. 4.2) but it is possible to obtain a local approximation to linearity by means of a Taylor's expansion in which second-order terms and above are ignored. Because of this the method was referred to as a first-order second moment method.

The method, reported by Cornell (1969) and Rosenblueth and Esteva (1972), consisted of expanding the limit state function at the mean point (i.e. mean values of the variables were inserted into the expression for Z), in order to create the local approximation to a linear failure boundary. Because of this the method was referred to as a mean value first-order second moment method.

THE ADVANCED FIRST-ORDER SECOND MOMENT METHOD

Although a significant step forward, Cornell's method had the serious disadvantage that the position of the boundary approximation could vary with the way the expression for Z was written. The method regarded an expression such as $Z = 2x^2 + 2xy$ as quite different to the expression $Z = 2x(x + y)$ and would produce a different value of β for each expression.

This sensitivity of the reliability index was pointed out by Ditlevsen (1973) and an invariant second moment index was proposed by Hasofer and Lind (1974). The authors showed that an invariant reliability index is obtained if the point chosen for the linear approximation is actually on the failure boundary. This point of maximum probability of failure lies somewhere along the boundary and is generally called the design point and given the symbol x^*.

Hasofer and Lind's work was extended by Rackwitz (1976) and led to the advanced first-order second moment method which will now be briefly described.

It has been established that the failure, i.e. the limit state, of a structure can be expressed as a function of the relevant basic variables:

$$Z = g(X) = g(X_1, X_2, X_3, \ldots, X_n)$$

If $g(X)$ consists of the single variable, X, then the failure boundary can be plotted (see Fig. 4.3).

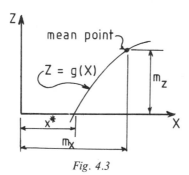

Fig. 4.3

From the plot it is seen that at x^* the value of Z (or $g(x^*)$) is zero. With more than one variable $g(x^*)$ is $g(x_1^*, x_2^*, x_3^*, \ldots, x_n^*)$ where $x_1^*, x_2^*, x_3^*, \ldots$ are the design values of X_1, X_2, X_3, etc.

Approximating the expression for Z as a linear expansion can be carried out by the following technique.

Consider first the case of Z equal to $g(X)$ where X is a single variable (Fig. 4.3). Then, using Taylor's expansion:

$$Z = g(x^*) + (x - x^*)g'(x^*) + \frac{(x - x^*)^2}{2} g''(x^*) + \cdots$$

where:

$x^* =$ the value of X at which the approximation is taken
$g'(x^*) = \mathrm{d}g(X)/\mathrm{d}X \mid x^* =$ the first derivative of $g(X)$ evaluated for $X = x^*$

Removing second-order terms and above results in a first-order approximation consisting of two terms:

$$Z = g(x^*) + (x - x^*)g'(x^*)$$

If Z is a function of several variables the equivalent expression is:

$$Z = g(x_1^*, x_2^*, x_3^*, \ldots, x_n^*) + \sum_{i=1}^{n} (x_i - x_i^*)g'(x_i^*)$$

$$= \sum_{i=1}^{n} (x_i - x_i^*)g(x_i^*)$$

as

$$g(x_1^*, x_2^*, x_3^*, \ldots, x_n^*) = 0$$

Therefore:

$$m_z \doteq g(m_1^*, m_2^*, m_3^*, \ldots, m_n^*) + \sum_{i=1}^{n} (m_i - x_i^*)g'(x_i^*)$$

$$= \sum_{i=1}^{n} (m_i - x_i^*)g'(x_i^*)$$

as

$$g(m_1^*, m_2^*, m_3^*, \ldots, m_n^*) = 0$$

and:

$$\sigma_z \doteq \sqrt{\sum_{i=1}^{n} [g'(x_i^*)\sigma_i]^2}$$

where $g'(x_i^*)$ is the first derivative of $g(X)$ evaluated at the point where x^* equals $(x_1^*, x_2^*, x_3^*, \ldots, x_n^*)$.

The sensitivity factor

A measure of the contribution of any variable, X_i, to the value of σ_z is its sensitivity factor, α_i, which is simply the ratio:

$$\alpha_i = \frac{g'(x_i^*)\sigma_i}{\sigma_z}$$

Now:

$$\sigma_z^2 \doteq \sum_{i=1}^{n} [g'(x_i^*)\sigma_i]^2$$

$$= \sum_{i=1}^{n} (\alpha_i \sigma_z)g'(x_i^*)\sigma_i$$

$$= \sigma_z \sum_{i=1}^{n} \alpha_i g'(x_i^*)\sigma_i$$

Hence:

$$\sigma_z \doteq \sum_{i=1}^{n} \alpha_i g'(x_i^*)\sigma_i$$

Now:

$$\beta = \frac{m_z}{\sigma_z} = \frac{\sum_{i=1}^{n} (m_i - x_i^*)g'(x_i^*)}{\sum_{i=1}^{n} \alpha_i g'(x_i^*)\sigma_i}$$

Therefore

$$\sum_{i=1}^{n} g'(x_i^*)[(m_i - x_i^*) - \alpha_i \beta \sigma_i] = 0$$

The value of x_i^* that satisfies this equation is given by:

$$x_i^* = m_i - \alpha_i \beta \sigma_i$$

for all values of i.

By determining all values of x_i^* the design point x^* can be obtained. The solution technique given in C.I.R.I.A.'s Report 63 (1976) is as follows:

(1) Guess a value for β
(2) Set $x_i^* = m_i$ for all i values
(3) Compute $\partial g/\partial x_i$ for all i, at $x = x^*$
(4) Compute α_i for all i
(5) Compute new x_i^* values
(6) Repeat steps 3 to 5 until stable values of x_i^* are achieved
(7) Evaluate $Z = g(x_1^*, x_2^*, x_3^*, \ldots, x_n^*)$
(8) Modify β and repeat steps 3 to 7 to achieve $Z = 0$

Example 4.2

Show that a value of 1.156 for β determines the design point of example 4.1

Solution

$$Z = \sigma \tan \phi - \tau = 0$$

$$= g(X_1, X_2, C)$$

where:

$X_1 = \sigma$
$X_2 = \phi$
$C = 50 \text{ kN/m}^2$

setting x_i^* values equal to m_i^* gives $x_1^* = 100 \text{ kN/m}^2$ and $x_2^* = 35°$.

$$g'(x_1^*) = \frac{\partial Z}{\partial X_1} = \tan \phi \mid x_1^*; x_2^* = 0.7002$$

$$g'(x_2^*) = \frac{\partial Z}{\partial X_2} = \sec^2 \phi \mid x_1^*; x_2^* = 149.03$$

With these values:

$$\sigma_Z = \sqrt{\left(\frac{\partial g}{\partial X_1}\sigma_{X1}\right)^2 + \left(\frac{\partial g}{\partial X_2}\sigma_{X2}\right)^2}$$

$$= \sqrt{(0.7007 \times 20)^2 + (149.03 \times 0.0873)^2}$$

$$= 19.1149$$

$$\alpha_1 = \frac{g'(x_1{}^*)\sigma_{X1}}{\sigma_Z} = \frac{0.7007 \times 20}{19.1149} = 0.7326$$

$$\alpha_2 = \frac{g'(x_2{}^*)\sigma_{X2}}{\sigma_Z} = \frac{149.03 \times 0.0873}{19.1149} = 0.6806$$

Therefore:

$$x_1{}^* = m_1 - \alpha_1\beta\sigma_{X1} = 100 - 0.7362 \times 1.156 \times 20 = 83.06$$

$$x_2{}^* = m_2 - \alpha_2\beta\sigma_{X2} = 0.6109 - 0.6806 \times 1.156 \times 0.0873$$

$$= 0.5422 \text{ rads}$$

$$= 31.06°$$

The full iteration is set out below:

Iteration	Variables	$g'(x_i{}^*)$	α_i	$x_i{}^*$
START	$X_1 = \sigma$			100 kN/m^2
	$X_2 = \phi$			$35°$
1	X_1	0.7002	0.7328	83.06
	X_2	149.03	0.6806	31.06
2	X_1	0.6024	0.7733	82.12
	X_2	113.20	0.6340	31.34
3	X_1	0.6089	0.7783	82.01
	X_2	112.56	0.6279	31.37
4	X_1	0.6097	0.7790	81.99
	X_2	112.49	0.6271	31.38
5	X_1	0.6098	0.7790	82.00
	X_2	112.48	0.6270	31.38

Using the final derived values for X_1 and X_2 the closing error for Z can be obtained and equals 0.0137, a value that most engineers would accept as equivalent to zero. Therefore: design point $= (82 \text{ kN/m}^2, 31.38°)$ and $\beta = 1.156$.

Reduced variables

It is generally more convenient to work in terms of 'reduced', 'normalised', or 'standardised' variables which were described in chapter 3.

If x_1 is the particular value of a variable with a mean of m_1 and a standard deviation of σ_1, then the corresponding reduced variable, y_1, is given by the expression:

$$y_1 = \frac{x_1 - m_1}{\sigma_1}$$

A reduced variable has the propertics that its mean value is zero and its standard deviation is one which means that the origin of the axes that represent this reduced space is also the mean point of the reduced variables. The failure surface can now be expressed as:

$$Z = h(y)$$

where:

$$h(y) = h(y_1, y_2, y_3, \ldots, y_n)$$

Taylor's first degree approximation to Z at the point x^* has already been established:

$$Z \doteqdot g(x_1{}^*, x_2{}^*, x_3{}^*, \ldots, x_n{}^*) + \sum_{i=1}^{n} (x_i - x_i{}^*)g'(x_i{}^*)$$

The linear approximation at the point $y^* = (y_1{}^*, y_2{}^*, y_3{}^*, \ldots, y_n{}^*)$ is therefore:

$$Z \doteqdot h(y_1{}^*, y_2{}^*, y_3{}^*, \ldots, y_n{}^*) + \sum_{i=1}^{n} (y_i - y_i{}^*)h'(y_i{}^*)$$

simplifying to:

$$Z = \sum_{i=1}^{n} (y_i - y_i{}^*)h'(y_i{}^*)$$

The mean of Z is therefore:

$$m_Z \doteqdot \sum_{i=1}^{n} (m_i - y_i{}^*)h'(y_i{}^*) = -\sum_{i=1}^{n} h'(y_i{}^*)$$

(as the mean of a standardised variable is zero) and:

$$\sigma_Z = \sum_{i=1}^{n} \alpha_i h'(y_i{}^*)\sigma_i$$

$$= \sum_{i=1}^{n} \alpha_i h'(y_i{}^*)$$

(as the variance of a standardised variable is one. Now:

$$\beta = \frac{m_Z}{\sigma_Z} = -\frac{\sum\limits_{i=1}^{n} y_i h'(y_i{}^*)}{\sum\limits_{i=1}^{n} \alpha_i h'(y_i{}^*)}$$

Therefore:

$$\sum_{i=1}^{n} h'(y_i{}^*)[-(y_i{}^* - \alpha_i \beta)] = 0$$

The solution, in terms of the standardised variables is therefore:

$$y^* = -\alpha_i \beta \qquad \text{for all } i$$

From the above equation it is seen that the distance from the origin to y^* is a measure of the reliability index. It can be obtained from the expression:

$$\beta = \sqrt{\sum_{i=1}^{n} y_i{}^2}$$

ITERATIVE PROCEDURE FOR DETERMINING β

An algorithm proposed by Fiessler (1980) can be used with reduced variables and gives the value of β after only one set of iterations. A suitable procedure is as follows:

(1) Determine an expression for $g(X)$.
(2) Evolve an expression for $h(y)$.
(3) Determine expressions for all first derivative of $h(y)$, $h_i{}'$.
(4) Set $y = 0$ and $\beta = 0$.
(5) Evaluate all $h_i{}'$ values.
(6) Evaluate $h(y)$.
(7) Evaluate standard deviation of Z from:

$$\sigma_Z = \sqrt{\sum (h_i{}')^2}$$

(8) Evaluate new values for y from:

$$y = -\frac{h_i{}'}{\sigma_Z}\left[\beta + \frac{h(y)}{\sigma_Z}\right]$$

(9) Evaluate:

$$\beta = \sqrt{\sum y_i{}^2}$$

(10) Repeat steps 5 to 9 until values converge.

(Note that in step 7 the term σ_{yi} has been included but it is equal to one as y_i is a standardised variable.)

Example 4.3

Example 4.2 will be recalculated using reduced variables.

Solution

$$Z = \sigma \tan \phi - \tau$$
$$= g(X_1, X_2, C)$$

where:

$$X_1 = \sigma$$
$$X_2 = \phi$$
$$C = 50 \text{ kN/m}^2$$

Hence:

$$Z = g(X) = X_1 . \tan(X_2) - 50$$

i.e. the basic variables are therefore:

	Mean	s.d.
X_1	100	20 (kN/m^2)
X_2	35	5 (degrees)

Now:

$$X_1 = \sigma_1 y_1 + m_1; \qquad X_2 = \sigma_2 y_2 + m_2$$

Hence:

$$Z = h(y) = (y_1\sigma_1 + m_1) \tan(y_2\sigma_2 + m_2) - 50$$

$$h_1' = \frac{\partial Z}{\partial y}\bigg| y_1 = \sigma_1 \tan(y_2\sigma_2 + m_2)$$

$$h_2' = \frac{\partial Z}{\partial y}\bigg| y_2 = (y_1\sigma_1 + m_1) \sec^2(y_2\sigma_2 + m_2)$$

The first three steps of the iteration procedure have now been carried out. The procedure continues:

Step 5 – With:

$$y_1 = y_2 = 0; \qquad \beta = 0$$

then:

$$h_1' = 14.004$$

$$h_2' = 13.005$$

Step 6

$$h(y) = 100 \times 0.7002 - 50 = 20.02$$

Step 7

$$\sigma_Z = \sqrt{14.004^2 + 13.005^2} = 19.112$$

Step 8

$$y_1 = \frac{-14.004}{19.112}\left[0 + \frac{20.02}{19.112}\right] = -0.767; \qquad y_2 = -0.713$$

Step 9

$$\beta = \sqrt{0.767^2 + 0.713^2} = 1.047$$

The procedure is now continued by returning to step 5 and inserting the derived values for y_1, y_2 and β where appropriate.

The complete iteration is set out below.

Iteration	y_1	y_2	β	$h(y)$
1	0.0	0.0	0.0	20.02
2	-0.767	-0.713	1.048	1.7416
3	-0.890	-0.740	1.157	-0.0177
4	-0.899	-0.727	1.156	-0.0018
5	-0.900	-0.725	1.156	-0.0000
6	-0.900	-0.725	1.156	-0.0000

The reliability index = 1.156.

It can be seen that the value of β is identical to that obtained in terms of the original variables, X_i.

The design point in terms of the standardised variables, y_i, is of course $(-0.9, -0.725)$ but the design point in terms of the original variables can also be obtained:

$$X_1 = y_1\sigma_1 + m_1 = -0.9 \times 20 + 100 = 82 \text{ kN/m}^2$$

$$X_2 = y_2\sigma_2 + m_2 = -0.725 \times 5 + 35 = 31.375°$$

For this two-dimensional problem a graphical solution is possible as the failure boundary can be plotted in a similar manner to that of example 4.1. By selecting a range of suitable values for y_1, determining X_1 and calculating the X_2 values, the corresponding set of y_2 values can be obtained.

y_1	−2.0	−1.5	−1.0	−0.5	0.0	0.5	1.0
X_1 (kN/m²)	60	70	80	90	100	110	120
X_2 (°)	39.8	35.5	32.0	29.1	26.6	24.4	22.6
y_2	0.96	0.1	−0.6	−1.18	−1.68	−2.12	−2.48

The failure boundary is shown plotted in transformed variable space in Fig. 4.4. It can be seen that the value of β scales 1.16.

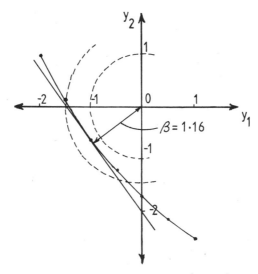

Fig. 4.4   Example 4.3

Example 4.4

A short column has a diameter X_1 and is loaded with an axial compressive load X_2. The ultimate compressive stress of the column is X_3. The variables have the following mean and s.d. values:

	Mean	s.d.
X_1	3.5	0.4
X_2	10.0	1.0
X_3	2.5	0.5

Determine the reliability index of the system.

Solution

As $Z = R - S$, the limit state function can be written as:

$$Z = g(X) = X_3 \frac{\pi}{4} X_1{}^2 - X_2 = 0 \qquad (1)$$

or as:

$$Z = g(X) = \frac{\pi}{4} X_1{}^2 - \frac{X_2}{X_3} = 0 \qquad (2)$$

or as:

$$Z = g(X) = \frac{X_3}{X_2} \frac{\pi}{4} - X_1{}^2 - 1 = 0 \qquad (3)$$

Expressing equation (1) in terms of reduced variables gives:

$$Z = h(y) = (y_3\sigma_3 + m_3)\frac{\pi}{4}(y_1\sigma_1 + m_1)^2 - (y_2\sigma_2 + m_2)$$

and the expression for the differentiations of Z with respect to y_1, y_2 and y_3 are:

$$h_1' = 2\sigma_1(y_3\sigma_3 + m_3)\frac{\pi}{4}(y_1\sigma_1 + m_1)$$

$$h_2' = -\sigma_2$$

$$h_3' = \sigma_3\frac{\pi}{4}(y_1\sigma_1 + m_1)^2$$

Starting with $y_1 = y_2 = y_3 = 0$ gives:

$$\beta = 0.00; \quad h(y) = 14.05; \quad h_1' = 5.498; \quad h_2' = -1.0; \quad h_3' = 4.811$$

Hence:

$$\sigma_Z = \sqrt{5.498^2 + 1^2 + 4.811^2} = 7.374$$

and, after the first iteration:

$$y_1 = \frac{-5.498}{7.374}\left[0 + \frac{14.05}{7.374}\right] = -1.421; \quad y_2 = 0.258; \quad y_3 = -1.243$$

which gives a β value of 1.90.

The full iteration procedure for β is set out below.

	y_1	y_2	y_3	β	$h(y)$
1	0.0	0.0	0.0	0.0	14.05
2	−1.42	0.26	−1.24	1.90	2.43
3	−1.68	0.48	−1.64	2.40	0.07.
4	−1.62	0.54	−1.71	2.42	−0.03
5	−1.59	0.54	−1.73	2.42	−0.01
6	−1.59	0.54	−1.74	2.42	0.00

The reliability index for the system = 2.42.

The reader might like to show the invariance of the reliability index by selecting one of the other two limit state equations and checking that the same value is obtained for β.

Determination of P_f

Civil engineers are familiar with quality control and will readily accept statements such as: 'One out of every hundred of these concrete slabs is liable to be bad' or 'There is a probability of failure of 1 in 100 with these slabs'. Most civil engineers can therefore accept the idea that for any structure there will be a certain risk of failure whilst few will appreciate the idea of a reliability index.

Provided that the variables involved have probability distributions that are close to normal and provided that the linear approximation of the failure surface is realistic, then an exact value for P_f can be obtained from the expression:

$$P_f = \Phi(-\beta)$$

where $\Phi(-\beta)$ is the general symbol for the value of the cumulative probability of Z (for $-\infty$ to $-\beta$). This value can be obtained from tables or from a suitably programmed microcomputer or calculator and is the area under the standardised normal density function, as illustrated in Fig. 4.5.

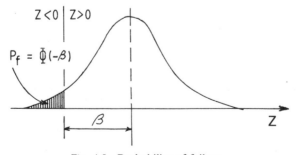

Fig. 4.5 Probability of failure

If the variables are not normal or if the linear approximation is poor the P_f value obtained from the above formula is referred to as the 'notional (or nominal), probability of failure'.

An important question is whether or not a linear approximation of the failure surface is accurate enough. Checks can be carried out by Monte Carlo and other simulation methods but a number of better approximations have been developed, Ditlevsen (1976), Fiessler et al. (1979), and these generally show very good agreement with first order reliability methods. These better methods use a quadratic expansion of $g(X) = 0$ and are called 'second order reliability methods'.

It can safely be assumed that, for almost all engineering problems, the linear approximation of the failure surface will be adequate and it is worth remembering that if there are several variables, of roughly equal weight, the resulting Z function tends to be normal, even when the separate variables are not themselves normal, Benjamin and Cornell (1970). Nevertheless, if it is known that some of the variables involved in the design are non-normal, then the accuracy of the determined value of P_f is improved if this information is incorporated into the reliability analysis. This can be achieved by a method proposed by Fiessler and Rackwitz (1976), and described in chapter 8.

However, it should be remembered that with geotechnical problems, due to the inevitable lack of statistical information, any P_f values obtained are nominal.

For interest, the nominal probability of failure for example 4.2 is found from appendix III using the expression:

$$P_f = \Phi(-\beta) = \Phi(-1.156) = 0.124 \qquad (\sim 12\%)$$

Chapter Five

Applications of the Second Moment Method

Structures

The application of the second moment method to structural problems is straightforward as the design parameters are also the basic variables of the limit state equation, Z.

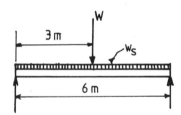

Fig. 5.1   Example 5.1

Example 5.1

Consider the simply supported beam shown in Fig. 5.1. The beam is of reinforced concrete of dead weight 5 kN/m run and spans 6 m. Both of these values may be regarded as constant.

The beam supports a uniform load w_s and a central point load, W, and has an ultimate moment of resistance M_f. The values of these variables are set out below.

Variable	Symbol	Mean value (m)	s.d. (σ)	Units
M_f	X_1	300	30	kN/m
w_s	X_2	30	3	kN/m
W	X_3	25	5	kN

Determine the reliability index against bending failure.

Solution
Resistive moment $= M_f = R$
Disturbing moments:

$$\text{due to dead weight} = \frac{5 \times 6^2}{8} = 22.5 \,\text{kN/m}$$

$$\text{due to } w_s \text{ and } W = \frac{w_s \times 6^2}{8} + \frac{6W}{4} = 4.5w_s + 1.5W$$

Hence total disturbing moment is $22.5 + 4.5w_s + 1.5W$ which equals S.
Now:

$$Z = R - S = M_f - 22.5 - 4.5w_s - 1.5W$$

or:

$$Z = g(X) = X_1 - 4.5X_2 - 1.5X_3 - 22.5$$

and expressing in reduced variables:

$$Z = h(y) = (y_1\sigma_1 + m_1) - 4.5(y_2\sigma_2 + m_2) - 1.5(y_3\sigma_3 + m_3) - 22.5$$

Differentiating and inserting mean values gives:

$$h_1' = \sigma_1 = 30; \qquad h_2' = -4.5\sigma_2 = -13.5; \qquad h_3 = -1.5\sigma_3 = -7.5$$

Hence:

$$\sigma_Z = \sqrt{(30^2 + 13.5^2 + 7.5^2)} = 33.74$$

and, after the first iteration:

$$y_1 = -2.767; \qquad y_2 = 1.245; \qquad y_3 = 0.692$$

Convergence is rapid and the full iteration for β is set out below.

	y_1	y_2	y_3	β	$h(y)$
1	0.0	0.0	0.0	0.0	105
2	-2.77	1.25	0.69	3.11	0
3	-2.77	1.25	0.69	3.11	0

The reliability index for bending $= 3.11$.

Example 5.2

An elastic cantilever beam is subjected to a point load, W, as shown in
Fig. 5.2. Ignoring the self weight of the beam and assuming that the
dimensions given in the figure are constant, determine the value of the

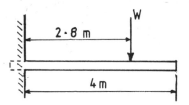

Fig. 5.2 Example 5.2

reliability index for the deflection at the free end of the cantilever not to exceed 100 mm. The relevant variables are listed below.

Variable	Symbol	Mean value	s.d.	Units
Elasticity Modulus (E)	X_1	20 000	2500	MN/m^2
Moment of Intertia (I)	X_2	0.01	0.0005	m^4
(W)	X_3	800	125	kN

Solution

Reynolds (1957) quotes the deflection at the free end of the cantilever as being equal to:

$$\frac{kWL^3}{EI}$$

where:

$$k = \frac{a^2(3-a)}{6} = 0.1878$$

a being the ratio of the dimension 2.8 m to the span of the cantilever and being equal to 0.7 in this example. Hence:

$$Z = 100 - \frac{0.1878WL^3}{EI}$$

(As E is expressed in MN/m^2 the second term of the expression is in millimetres.), i.e.

$$Z = 100EI - 0.1878WL^3 = 100X_1.X_2 - 12.0192X_3$$
$$= 100(y_1\sigma_1 + m_1)(y_2\sigma_2 + m_2) - 12.0192(y_3\sigma_3 + m_3)$$

The full iterative procedure for β gives the following values:

	y_1	y_2	y_3	β	$h(y)$
1	0.0	0.0	0.0	0.0	10 384.6
2	−2.45	−0.98	1.23	2.91	1724.1
3	−2.94	−0.86	1.55	3.43	164.7
4	−3.00	−0.79	1.57	3.48	18.8
5	−3.01	−0.78	1.57	3.49	0.53
6	−3.02	−0.78	1.57	3.49	0.09
7	−3.02	−0.78	1.57	3.49	0.01

Example 5.3

Determine the reliability index for example 5.2 if the dimension 2.8 m has a coefficient of variation of 7.5%.

Solution

$$k = \frac{a^2(3 - a)}{6}$$

and, with the mean value 2.8 m, k equals 0.1878.
   Standard deviation of the dimension equals $0.075 \times 2.8 = 0.21$ m.
   If the dimension is one standard deviation greater than 2.8 m then it equals 3.01 m and k equals 0.2121 whereas if the dimension is one standard deviation less than 2.8 m it is equal to 2.59 and k equals 0.1644. An approximate value for the s.d. of k is therefore 0.5 (0.2121 − 0.1644) = 0.0239. (This approximate method for determining the standard deviation of a variable is described later in this chapter.)
   Hence the relevant basic variables are:

Variable	Symbol	Mean value	s.d.	Units
E	X_1	20 000	2500	MN/m^2
I	X_2	0.01	0.0005	m^4
W	X_3	800	125	kN
k	X_4	0.1878	0.0239	

The expression for Z is now:

$$Z = 100(y_1\sigma_1 + m_1)(y_2\sigma_2 + m_2) - 64(y_3\sigma_3 + m_3)(y_4\sigma_4 + m_4)$$

and the iterative procedure for β is as set out below:

	y_1	y_2	y_3	y_4	β	$h(y)$
1	0.0	0.0	0.0	0.0	0.0	10384.6
2	−2.36	−0.94	1.42	1.16	3.13	−34.74
3	−2.20	−0.65	1.59	1.38	3.12	−105.85
4	−2.17	−0.65	1.59	1.37	3.09	−0.12
5	−2.17	−0.65	1.59	1.37	3.09	−0.01

The reliability index is reduced to 3.09.

Example 5.4

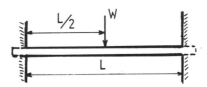

Fig. 5.3   Example 5.4

A beam, fixed at both ends, has a span, L, of 5 m and is loaded with a central point load W as shown in Fig. 5.3.

The central deflection of the beam is equal to:

$$\frac{1}{192}\frac{WL^3}{EI}$$

Determine the reliability index that the central deflection will not be greater than $L/100$ given that L is of fixed value and that the values of the relevant variables are:

Variable	Symbol	Mean value	s.d.	Units
W	X_1	70	8	kN
E	X_2	22 000 000	4 500 000	kN/m²
I	X_3	0.00015	0.00002	m⁴

$$R = \text{maximum allowable deflection} = \frac{5}{100} = 0.05 \text{ m}$$

$$S = \text{deflection at centre of span} = \frac{1}{192}\frac{W\times 5^3}{EI} = 0.6510\frac{W}{EI}$$

Hence:

$$Z = h(y) = 0.05(y_2\sigma_2 + m_2)(y_3\sigma_3 + m_3) - 0.651(y_1\sigma_1 + m_1)$$

The iterative procedure for β gives:

	y_1	y_2	y_3	β	$h(y)$
1	0.0	0.0	0.0	0.0	119.43
2	0.38	-2.44	-1.59	2.94	17.50
3	0.63	-3.22	-1.33	3.54	2.45
4	0.62	-3.28	-0.89	3.46	-0.95
5	0.57	-3.28	-0.80	3.42	-0.04
6	0.57	-3.28	-0.79	3.42	-0.00
7	0.57	-3.28	-0.79	3.42	-0.00

Example 5.5

Determine the reliability index for example 5.4 if the dimension L has a mean value of 5 m and a coefficient of variation of 5%.

Solution
The basic variables are now:

Variable	Symbol	Mean value	s.d.	Units
W	X_1	70	8	kN
E	X_2	22 000 000	4 500 000	kN/m^2
I	X_3	0.00015	0.00002	m^4
L	X_4	5	0.25	m

The expression for Z becomes:

$$Z = g(X) = \frac{X_4 X_2 X_3}{100} - \frac{X_1 X_4{}^3}{192}$$

$$= h(y) = ((y_4\sigma_4 + m_4)(y_2\sigma_2 + m_2)(y_3\sigma_3 + m_3)/100$$
$$- 0.00521(y_1\sigma_1 + m_1)(y_4\sigma_4 + m_4)^3$$

The expression for h_4' is the only difficult differentiation:

$$h_4' = \sigma_4(y_2\sigma_2 + m_2)(y_3\sigma_3 + m_3) - 0.0156\sigma_4(y_1\sigma_1 + m_1)(y_4\sigma_4 + m_4)^2$$

The iterative procedure for β gives:

	y_1	y_2	y_3	y_4	β	$h(y)$
1	0.0	0.0	0.0	0.0	0.0	119.41
2	0.38	-2.44	-1.59	-0.10	2.94	18.00
3	0.62	-3.20	-1.33	0.46	3.55	-4.32
4	0.62	-3.18	-0.87	0.59	3.41	-0.87
5	0.59	-3.18	-0.82	0.54	3.38	-0.03
6	0.58	-3.18	-0.81	0.53	3.38	-0.00

The reliability index has reduced to 3.38.

Soils

Most soil mechanics limit state functions contain very few basic variables. For example the strength of a soil structure involves three variables only: unit weight, γ, cohesion, c, and the angle of shearing resistance, ϕ. The difficulty is that there are often terms which are functions of ϕ. Examples:

$$K_a = \frac{1 - \sin\phi}{1 + \sin\phi}; \qquad \mu = \tan\phi; \qquad \text{etc.}$$

As an illustration of the complications that can arise consider the relatively simple problem of a concentric column load supported by a square footing, of dimensions $B \times B \text{ m}^2$ and founded at a depth of z metres. The accepted expression for the ultimate load, Q_u, is:

$$Q_u = (1.3cN_c + \gamma z N_q + 0.4\gamma B N_\gamma)B^2$$

N_c, N_q and N_γ are bearing capacity coefficients and the factor of safety of the foundation against bearing capacity failure is P/Q_u.

Now, N_c, N_q and N_γ are all functions of ϕ and if Hansen's expressions, described later in this chapter, are adopted then the limit state function for the bearing capacity failure of the foundation can be written down:

$$Z = B^2 c\left[\tan^2\left(45° + \frac{\phi}{2}\right)\exp(\pi\tan\phi) - 1\right]\cot\phi$$

$$+ \gamma z B^2 \tan^2\left(45° + \frac{\phi}{2}\right)\exp(\pi\tan\phi)$$

$$+ \gamma B^3 \tan\phi\left[\tan^2\left(45° + \frac{\phi}{2}\right)\exp(\pi\tan\phi) - 1\right] - P - W$$

where W is the weight of the column and foundation less the weight of the excavated soil.

In order to use the second moment approach this expression must be converted into reduced variables and then differentiated. When it is considered that the average bearing capacity problem will include extra terms such as inclined load factors, eccentricity of loading, etc., it is apparent that there will be many cases in geotechnics where a limit state function becomes unmanageable and that some form of simplification must take place if the second moment approach is to be used.

A simplified approach, suggested by the author, Smith (1986), in which the various functions of ϕ are regarded as forming a set of independent variables, each with its own expected (or mean) value and its own standard deviation, will be used here.

TREATMENT OF FUNCTIONS OF ϕ

It can be shown that (Benjamin and Cornell (1970)) if:

$$U = g(X)$$

where $X = (X_1, X_2, X_3, \ldots, X_n)$, then:

$$\sigma_U = \sqrt{\sum_{i=1}^{n} (g_i' \sigma_X)^2}$$

where:

$$g_i' = \frac{\partial U}{\partial X}\bigg|_{m_i}$$

i.e. the first derivative of U with $X = m_i$. If X is a single variable then:

$$\sigma_U = \sqrt{\frac{\partial U}{\partial X}\bigg| (m_X \cdot \sigma_X)^2}$$

Example 5.6

The angle of friction of a soil has a mean value of 35° and a standard deviation of 5°. Determine the mean value and standard deviation of $\tan \phi$.

Solution
Let:

$$Y = \tan \phi$$

Then:

$$m_y = \tan \phi \,|\, m_\phi = 0.7002; \qquad g_\phi' = \frac{\partial Y}{\partial \phi} = \sec^2 \phi \,|\, m_\phi = 1.490$$

$$\sigma_y = \sqrt{(g_\phi' \sigma_\phi)^2} = \sqrt{(1.49 \times 0.0873)^2} = 0.1301$$

Example 5.7

The Rankine expression for the coefficient of active earth pressure is:

$$K_a = \frac{1 - \sin \phi}{1 + \sin \phi}$$

Determine the mean value and standard deviation of K_a for a soil whose angle of friction has a mean value of 33° and a standard deviation of 2°.

Solution
Mean value of:

$$K_a = \frac{1 - \sin \phi}{1 + \sin \phi} \Bigg|_{m^\phi} = \frac{1 - \sin 33°}{1 + \sin 33°} = 0.2948$$

Now:

$$K_a = g(\phi)$$

and:

$$\sigma_{K_a} = \sqrt{\left(\frac{\partial K_a}{\partial \phi} m_\phi \cdot \sigma_\phi\right)^2}$$

$$\frac{\partial K_a}{\partial \phi}\Bigg|_{m_\phi} = \frac{-2 \cos \phi}{(1 + \sin \phi)^2}\Bigg|_{m_\phi} = \frac{-2 \cos 33}{(1 + \sin 33)^2} = -0.703$$

Standard deviation of:

$$K_a = \sqrt{\left(-0.703 \times \frac{2\pi}{180}\right)^2} = 0.0245$$

Treatment of bearing capacity factors

Meyerhof's equations (1955) for the bearing capacity coefficients N_c and N_q are now generally used in geotechnics as they are recognised as being probably the most satisfactory.

$$N_c = (N_q - 1) \cot \phi; \qquad N_q = \tan^2\left(45° + \frac{\phi}{2}\right) \exp(\pi \tan \phi)$$

Unfortunately there is not the same firmness of opinion about the remaining factor, N_y. For this text the writer decided to use Hansen's equation (1970):

$$N_y = 1.5(N_q - 1) \tan \phi$$

i.e.

$$N_y = 1.5 \tan \phi \left[\tan^2\left(45° + \frac{\phi}{2}\right) \exp(\pi \tan \phi)\right] - 1.5 \tan \phi$$

The mean value of N_y can be quickly found by inserting the mean values of ϕ into the above equation.

The standard deviation of N_y can be found from the expression:

$$\sigma_{N_y} = \frac{\partial N_y}{\partial \phi}\Bigg|_{m_\phi} \cdot \sigma_\phi$$

involving the differentiation of the equation for N_y which, although tedious, is relatively simple and leads to the expression:

$$\frac{\partial N_y}{\partial \phi} = 1.5 \tan \phi \left[\frac{2 \cos \phi}{(1 - \sin \phi)^2} \exp(\pi \tan \phi) \right.$$

$$+ \tan^2\left(45° + \frac{\phi}{2}\right)\pi \sec^2 \phi(\exp \pi \tan \phi) \Bigg]$$

$$+ 1.5 \sec^2 \phi \left[\tan^2\left(45° + \frac{\phi}{2}\right) \exp(\pi \tan \phi) - 1 \right]$$

Values of N_c, N_q, N_y and their derivatives are given in appendices V, VI and VII.

Example 5.8

A granular soil has an angle of friction with a mean value of 40° and a coefficient of variation of 2.5%. Determine the corresponding mean and standard deviations values for the bearing capacity coefficient, N_y.

Solution

$$V_\phi = 0.025$$

Therefore:

$$\sigma_\phi = 0.025 \times 40 = 1° = 0.01745 \text{ radians}$$

From appendix VII, for $\phi = 40°$:

$$\text{mean } N_y = 79.54; \qquad \sigma_{N_y} = 805.05 \times 0.01745 = 14.051$$

Determination of s.d. values without differentiation

Some of the ϕ functions used in geotechnics are fairly complicated and their differentiation can present problems.

A way around this difficulty is to determine the values of the function for ϕ values one standard deviation on either side of the mean value of ϕ. The standard deviation of the function is then approximately equal to half of the difference between the two values.

Example 5.9

(a) Determine the standard deviation of K_a if ϕ has a mean value of 33° and a standard deviation of 2° (as in example 5.7).

(b) Determine the standard deviation of N_y if ϕ has a mean value of 40° and a standard deviation of 1° (as in example 5.8).

Solution
(a) ϕ values one standard deviation on either side of m_ϕ are 31° and 35°. Inserting these values into the formula for K_a gives K_a values of 0.3021 and 0.2710 respectively. Therefore:

$$\text{standard deviation of } K_a = \frac{0.3021 - 0.2781}{2} = 0.0245$$

(b) ϕ values one standard deviation on either side of m_ϕ are 39° and 41°. From appendix VII the corresponding N_y values are 66.76 and 95.05. Therefore:

$$\text{standard deviation of } N_y = \frac{95.05 - 66.76}{2} = 14.145$$

Example 5.10

Example 4.2 will now be recalculated using the simplified method.
Solution

$$Z = \sigma \tan \phi - \tau$$

or

$$Z = g(X_1, X_2, C)$$

where:

$$X_1 = \sigma; \quad X_2 = \tan \phi; \quad C = \tau$$

Hence:

$$Z = g(x) = X_1 . X_2 - 50$$

i.e. the basic variables are therefore:

	Mean	s.d.
X_1	100	20
X_2 (from example 5.6)	0.7002	0.1301

Now:

$$X_1 = \sigma_1 y_1 + m_1; \quad X_2 = \sigma_2 y_2 + m_2$$

Hence:

$$h(y) = (y_1\sigma_1 + m_1)(y_2\sigma_2 + m_2) - 50$$

$$h_1' = \frac{\partial Z}{\partial y}\bigg| y_1 = \sigma_1(y_2\sigma_2 + m_2)$$

$$h_2' = \frac{\partial Z}{\partial y}\bigg| y_2 = \sigma_2(y_1\sigma_1 + m_1)$$

The first three steps of the iteration procedure have now been carried out. The procedure continues:

Step 5   With $y_1 = y_2 = 0$ and $\beta = 0$ then:

$$h_1' = 14.004; \qquad h_2' = 13.01$$

Step 6

$$h(y) = 100 \times 0.7002 - 50 = 20.02$$

Step 7

$$\sigma_z = \sqrt{14.004^2 + 13.01^2} = 19.115$$

Step 8

$$y_1 = -\frac{14.004}{19.115}\left[0 + \frac{20.02}{19.115}\right] = -0.767; \qquad y_2 = -0.713$$

Step 9

$$\beta = \sqrt{0.767^2 + 0.713^2} = 1.047$$

The procedure is now continued by returning to step 5 and inserting the derived values for y_1, y_2 and β where appropriate.

   The complete iteration is set out below.

Iteration	y_1	y_2	β	$h(y)$
1	0.0	0.0	0.0	20.02
2	-0.767	-0.713	1.048	1.7416
3	-0.890	-0.740	1.157	-0.0177
4	-0.899	-0.727	1.156	-0.0018
5	-0.900	-0.725	1.156	-0.0000
6	-0.844	-0.759	1.135	-0.0000

The reliability index = 1.135.

As described in the introduction the nominal probability of failure can be found from the expression $P_f = \Phi(-\beta)$ using appendix III. Hence:

$$P_f = \Phi(-\beta) = \Phi(-1.135) = 0.128(\sim 12.8\%)$$

Example 5.10 is an interesting test of the simplified method in that the exact value of $\beta(1.156)$ is known from the work described in chapter 4. The simplified method, for this example, is in error by less than 2%.

Example 5.11
A surface reinforced concrete strip foundation, unit weight 24 kN/m³, is 2 m wide, 0.5 m thick and will be subjected to a uniform normal pressure, p, of mean value 500 kN/m² and coefficient of variation, V_p, of 6%.
   The soil is cohesionless.

Unit weight: mean $= 18$ kN/m³; $V_\gamma = 5\%$
Angle of friction: mean $= 40°$; $V_\phi = 2.5\%$

Determine the reliability index against bearing capacity failure.

Solution
The ultimate bearing capacity, q_u, of a surface strip foundation resting on cohesionless soil can be found from the expression:

$$q_u = 0.5B\gamma N_\gamma$$

For this example the width B can be regarded as constant at 2 m and the expression becomes:

$$q_u = \gamma N_\gamma \quad (=R)$$

S consists of two parts, the applied pressure, p, which can vary, and the weight of the foundation, $0.5 \times 24 = 12$ kN/m², which can be assumed to be constant. Now:

$$Z = R - S$$

$$= q_u - p - 12$$

$$V_\gamma = 5\% \text{ so, } \sigma_\gamma = 18 \times 0.05 = 0.9 \text{ kN/m}^3$$

$$V_p = 6\% \text{ so, } \sigma_p = 500 \times 0.06 = 30 \text{ kN/m}^2$$

and, from example 5.8, mean $N_\gamma = 79.54$ and $\sigma_{N_\gamma} = 14.05$.
   There are therefore three variables involved:

Parameter	Variable	Mean value	Standard deviation
γ (kN/m³)	X_1	18	0.9
N_γ	X_2	79.54	14.05
p (kN/m²)	X_3	500	30

i.e.

$$Z = g(X) = X_1 . X_2 - X_3 - 12$$
$$= h(y) = (y_1\sigma_1 + m_1)(y_2\sigma_2 + m_2) - (y_3\sigma_3 + m_3) - 12$$
$$h_1' = \sigma_1(y_2\sigma_2 + m_2)$$
$$h_2' = \sigma_2(y_1\sigma_1 + m_1)$$
$$h_3' = -\sigma_3$$

Using the suggested iterative procedure gives the following:

Iteration	y_1	y_2	y_3	β	$h(y)$
1	0.0	0.0	0.0	0.0	919.72
2	−0.94	−3.32	0.39	3.48	39.53
3	−0.44	−3.58	0.45	3.64	−11.59
4	−0.38	−3.55	0.43	3.59	−0.09
5	−0.38	−3.55	0.43	3.59	0.00

Reliability index = 3.59.
Nominal $P_f = 2.0 \times 10^{-4}$ (from appendix III).

Example 5.12

Figure 5.4 shows details of a mass stone rubble retaining wall which has a unit weight of 20 kN/m³. The retained soil has a level surface and carries a uniform surcharge, w_s with a mean value of 15 kN/m² and a standard deviation of 2.5 kN/m².

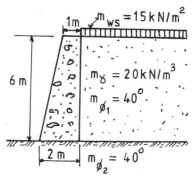

Fig. 5.4   Example 5.12

The retained soil is granular with the following properties:

Unit weight: mean value = 20 kN/m³;   s.d. = 1.5 kN/m³
Angle of friction: mean value = 40°;   s.d. = 1.5°

The foundation soil is granular and its angle of friction has a mean value of 40° and a standard deviation of 3°.

The coefficient of friction of the base of the wall and the foundation soil, μ, can be assumed to be equal to the tangent of the angle of friction of the foundation soil. Assuming that the back of the wall is smooth and using Rankine's formula for K_a, determine the reliability index against sliding.

Solution
The self weight of a structure is not generally treated as a basic variable (see chapter 6) and if we assume that the weight of the wall, W, is constant then:

$$W = 6 \times 20 \times \frac{(2+1)}{2} = 180 \text{ kN}$$

The total horizontal thrust from soil on to back of wall, S, is given by:

$$S = 6 . K_a . w_s + \tfrac{1}{2} . \gamma . K_a . 6^2 = 6K_a . w_s + 18 . \gamma . K_a$$

R, the sliding resistance is given by

$$W . \mu = 180$$

Therefore:

$$Z = 180\mu - 6K_a . w_s - 18K_a\gamma$$

There are four basic variables, γ, K_a, w_s and μ and, by using the methods of the previous examples, their means and standard deviations can quickly be found.

Variable	Symbol	Mean	s.d.	Units
γ	X_1	20	1.5	kN/m^3
K_a	X_2	0.2174	0.0149	
w_s	X_3	15	2.5	kN/m^2
μ	X_4	0.8391	0.0893	

Hence:

$$\begin{aligned} Z &= 180X_4 - 6X_2 . X_3 - 18X_1 . X_2 \\ &= 180(y_4\sigma_4 + m_4) - 6(y_2\sigma_2 + m_2)(y_3\sigma_3 + m_3) \\ &\quad - 18(y_1\sigma_1 + m_1)(y_2\sigma_2 + m_2) \end{aligned}$$

$$h_1' = -18\sigma_1(y_2\sigma_2 + m_2)$$

$$h_2' = -6\sigma_2(y_3\sigma_3 + m_3) - 18\sigma_2(y_1\sigma_1 + m_1)$$

$$h_3' = -6\sigma_3(y_2\sigma_2 + m_2)$$

$$h_4' = 180\sigma_4$$

The iterative procedure for β gives:

Iteration	y_1	y_2	y_3	y_4	β	$h(y)$
1	0.0	0.0	0.0	0.0	0.0	53.21
2	0.90	1.02	0.50	-2.46	2.85	-0.48
3	0.93	1.07	0.52	-2.39	2.83	-0.03
4	0.94	1.07	0.52	-2.39	2.82	-0.00
5	0.94	1.07	0.52	-2.39	2.82	0.00 —

The reliability index $= 2.82$.

The Monte Carlo simulation method

The use of numerical integration methods, such as the simplified level II method used to solve examples 5.2 to 5.8, is usually essential in civil engineering problems. Often the limit state equation contains functions of the basic variables that are too complicated for calculus to be used in the evaluation of their integrals. Another form of solution is that of simulation where the frequency distribution of Z is found by the use of random numbers. A particular form of this method is the Monte Carlo simulation which, with the advent of computers, has become increasingly popular. It can be extremely useful for the solution of complicated limit state equations and is described here.

A Monte Carlo simulation is a technique in which an output of random numbers is related to an assumed probability distribution (generally normal but other distributions may be used) so that a set of

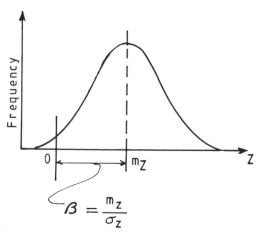

Fig. 5.5   Determination of β

probable values for the basic variables of the Z function can be obtained. With these values the corresponding value of Z (equal to R minus S) is found. The procedure is repeated many times and, if enough iterations are carried out then a histogram, and hence a frequency curve, of Z such as illustrated in Fig. 5.5, can be obtained.

Tables of random numbers are readily available but, nowadays, random numbers are usually generated by a computer.

The probability of failure can be expressed as the distance from the mean value of Z to the failure point (the point where Z equals zero). If this distance is expressed as $\beta\sigma_Z$ then we obtain the value of β:

$$\beta = \frac{m_Z}{\sigma_Z}$$

If desired it is possible to remove the possibility of outliers and to make certain that ϕ values do not reduce below ϕ_{cv} values. Cut-off points are inserted so that the generated maximum and minimum values of the basic variables are fixed at values that have less than some particular probability of occurrence. Usually a value of 5% is used. Then, for a normally distributed variable:

maximum value = mean value + 1.645 s.d.
minimum value = mean value − 1.645 s.d.

The first step in the procedure of determining random values of a particular variable is to first establish its probability distribution. If this is unknown then it is usually assumed to have a normal distribution.

If the 5% limitation is used then the value of the variable will be $m \pm k\sigma$ where k is either -1.645 or $+1.645$.

As established, the probability of k equalling -1.645 is 5% and the probability of k equalling $+1.645$ is 95%.

Hence, if the computer is programmed to produce 10 000 random numbers then a probability of 5% can be represented by the computer producing a number equal to or less than 500 whereas a probability of 95% can be represented by the computer producing a number equal to or greater than 9500.

If we adopt a cell width, k, of 0.25 then the range of k varies from -1.75 to $+1.75$ and there will be a total of 14 cells.

As an example the cell at the negative end of the distribution has k values varying from -1.75 to -1.5. From appendix III it is found that the probability of k equalling -1.75 is 0.0401 and the probability of k equalling -1.5 is 0.0668. Hence if the computer produces a random number between 401 and 668 we say that $k = -1.625$. If the number produced is less than 401 then k is taken as -1.645 (as this is the agreed cut-off point) although some people might argue that in this case k should be taken as equal to -1.875 (the central value of the next, imaginary, cell).

The full range of the central values of k for a normally distributed variable is set out below.

Random number	k
<401	-1.645
401 to 668	-1.625
668 to 1056	-1.375
1056 to 1587	-1.125
1587 to 2266	-0.875
2266 to 3085	-0.625
3085 to 4013	-0.375
4013 to 5000	-0.125
5000 to 5987	0.125
5987 to 6915	0.375
6915 to 7734	0.625
7734 to 8413	0.875
8413 to 8944	1.125
8944 to 9332	1.375
9332 to 9599	1.625
>9599	1.645

Each basic variable is treated as being independent.

Grigoriu (1983) maintains that if P_f is 10^{-4} then the number of iterations should be in the order of 10^5. In fact in the last three examples, and in the others of later chapters that were also checked, a stable value for β was established after some 5000 iterations. In most cases 10 000 iterations were carried out.

The values of β obtained by the two methods for the three examples are shown below.

	Level II method	Monte Carlo
Example 5.10	1.14	1.15
Example 5.11	3.59	3.80
Example 5.12	2.82	2.64

Chapter Six
More Probability Distributions

As has been indicated in the earlier chapters, the forms of the distributions of the basic variables in the limit state function have an effect on the accuracy of β and hence on the prediction of the probability of failure.

In structural engineering, provided an adequate test programme is carried out, the designer should be able to obtain a reasonable amount of statistical information about the constructional materials that he intends to use.

However, in soils engineering the statistical information that the designer requires if he is to accurately determine the various distributions of the soil parameters will generally not be available.

If the soils engineer has worked for some time in the area he may have some a priori knowledge, in the form of experience with similar soils or even results from similar work on a nearby site. With such knowledge he should be able to at least make meaningful assumptions as to the soil distributions.

For most practical situations the variability of a civil engineering parameter can be adequately described by one of three distributions.

(i) The normal distribution
(ii) The lognormal distribution
(iii) The beta distribution

The normal distribution has already been described in chapter 3 and, for the benefit of readers unfamiliar with the remaining two distributions, a few brief notes and examples now follow.

The lognormal distribution

A variable is said to have a lognormal (or logarithmicnormal) distribution when the logarithms of its values are normally distributed. We

137

can consider this distribution mathematically if we think in terms of two variables, X and Y, such that $Y = \exp(X)$.

If the relationship between the values of two variables is of the form:

$$Y = g(X)$$

and if y increases as x increases and if there is only a single value of y corresponding to each value of x, and vice versa, then we say that y is a monotonically increasing function with x.

Generally, if we know the function $Y = g(X)$ we can find the inverse function:

$$X = g^{-1}(Y)$$

In the case of a monotonically increasing function we can solve directly for the cdf of Y as the probability that Y is equal to or less than a particular value, y, must be equal to the probability that X is equal to or less than the value x, where x is the value corresponding to y, i.e. $g^{-1}(y)$. Hence:

$$F_Y(y) = P[Y \leqslant y] = P[X \leqslant x] = P[X \leqslant g^{-1}(y)] = F_X(g^{-1}(y))$$

Now, the pdf of Y can be obtained by differentiating its cdf:

$$f_Y(y) = \frac{d}{dy} F_Y(y) = \frac{d}{dy} \quad [F_X(g^{-1}(y))] = \frac{d}{dy} \int_{-\infty}^{g^{-1}(y)} f_X(x)\, dx$$

which simplifies to:

$$f_Y(y) = \frac{dg^{-1}}{dy}(y) f_X(g^{-1}(y))$$

Putting $x = g^{-1}(y)$

$$f_Y(y) = \frac{dx}{dy} f_X(x) \qquad \text{(A)}$$

In the case of a longnormal distribution:

$$X = \ln Y$$

$$Y = \exp(X)$$

Hence we can say that:

$$Y = g(X) = \exp(X)$$

$$X = g^{-1}(Y) = \ln Y \qquad \text{(normally distributed)}$$

and:

$$\frac{dx}{dy} = \frac{1}{y}$$

As X is normally distributed we can write down:

$$f_X(x) = \frac{1}{\sigma_X \sqrt{2\pi}} \exp\left[-\frac{1}{2}\left(\frac{x - m_x}{\sigma_X}\right)^2 \right] \qquad -\infty \leqslant x \leqslant \infty$$

And, with the relationship of expression (A) and substituting gives:

$$f_Y(y) = \frac{1}{y\sigma_X \sqrt{2\pi}} \exp\left[-\frac{1}{2}\left(\frac{\ln(y) - m_x}{\sigma_X}\right)^2 \right] \qquad y \geqslant 0$$

Note that X cannot have a negative value as the expression $X = \ln(-y)$ is meaningless.

It will be seen that the expression just derived for $f_Y(y)$ is in terms of m_X and σ_X, the mean and standard deviation of X, the logarithms of the values of Y. The expression can be improved if we make a substitution for $(\ln(y) - m_X)$. To enable us to do this we must think in terms of the medians of X and of Y.

The median of a variable, given the symbol $\breve{m}$, has been discussed in chapter 2 and is simply the middle value of the distribution when the values are placed in ranking order. If there are an even number of values then the median value is taken to be the average of the two central values.

It is obvious that the probability of a variable X having a value equal to or less than its median, $F_X(\breve{m}_X)$, is 0.5. Now, for any value y, the corresponding value of x will be $\ln(y)$. Hence:

$$P[X \leqslant \breve{m}_X] = P[X \leqslant \ln(\breve{m}_Y)] = 0.5$$

and:

$$P[X \leqslant \breve{m}_X] = 0.5$$

Hence:

$$\ln(\breve{m}_Y) = \breve{m}_X$$

For a normal, or indeed any symmetrical, distribution $\breve{m}_X$ equals m_X. Therefore:

$$\ln(\breve{m}_Y) = m_X$$

Hence:

$$(\ln(y) - m_X) = \ln(y) - \ln(\breve{m}_y)$$

Substituting in the expression for $f_Y(y)$:

$$f_Y(y) = \frac{1}{y\sigma_X\sqrt{2\pi}} \exp\left[-0.5\left(\frac{\ln(y) - \ln(\tilde{m}_y)}{\sigma_X}\right)^2\right] \qquad y \geqslant 0$$

Note: It can be shown that:

$$\sigma_X^2 = \ln(V_Y^2 + 1)$$

where V_Y is the coefficient of variation of $Y = \sigma_Y/m_y$, and:

$$\tilde{m}_Y = m_Y \exp(-0.5\sigma_X^2)$$

Example 6.1

Using the same mean and standard deviation values of example 3.10 determine the probability that X will lie between 40 and 50 kN/m², assuming a lognormal distribution.

Solution
It is perhaps useful to keep the symbol for the variable as Y to serve as a reminder that the variable is related.

$$V_Y = \frac{\sigma_Y}{m_Y} = \frac{6.56}{44.85} = 0.1463$$

$$\sigma_X^2 = \ln(0.1463^2 + 1) = 0.0212; \quad \Rightarrow \sigma_X = 0.1455$$

$$\tilde{m}_Y = 44.85 \exp(-0.5 \times 0.0212) = 44.38 \text{ kN/m}^2$$

Therefore:

$$f_Y(y) = \frac{1}{0.1445y\sqrt{2\pi}} \exp\left[-0.5\left(\frac{\ln(y) - \ln 44.38}{0.1455}\right)^2\right]$$

The appropriate values of $f_Y(y)$ corresponding to y are:

y	40	41	42	43	44	45	46	47	48	49	50
$f_Y(y)$	0.0533	0.0579	0.0611	0.0629	0.0626	0.0611	0.0582	0.0543	0.0496	0.0446	0.0393

and, using Simpson's rule, the area under the pdf curve is 0.560.

Alternative solution
Normal distribution information can be used to solve this problem. We have already established, for a lognormal distribution, the relationship between Y and X:

$$f_Y(y) = \frac{dx}{dy} f_X(x)$$

If we express $f_Y(y)$ in terms of its standardised variable then:

$$f_Y(y) = \frac{dz}{dy} f_Z(z)$$

where:

$$z = \frac{\ln(y) - \ln(\tilde{m}_y)}{\sigma_X}$$

Now:

$$\frac{dz}{dy} = \frac{1}{y} \cdot \frac{1}{\sigma_X}$$

Therefore:

$$f_Y(y) = \frac{f_Z(z)}{y\sigma_X}$$

Tables for the normal distribution can now be used to determine values of $f_Y(y)$. For example, for $f_Y(45)$:

$$z = (\ln 45 - \ln 44.38)/0.1455 = 0.01386/0.1455 = 0.0953$$

From Appendix II, for $z = 0.0953$, $f_Z(z)$ equals 0.3972. Therefore:

$$f_Y(y) = \frac{0.3972}{45 \times 0.1455} = 0.0607$$

However, the main advantage of using Z is in the determination of $F_Y(y)$ values, as $F_Y(y)$ equals $F_Z(z)$.

$$z_{40} = -0.7142$$

From Appendix III:

$$F_Z(-0.7142) = 0.2375$$

$$z_{50} = 0.8195$$

From Appendix III:

$$F_Z(0.8195) = 0.7937$$

Hence:

$$P[40 \leqslant Y \leqslant 50] = 0.7937 - 0.2375 = 0.556$$

The beta distribution

The pdf of this distribution, in its general form, is:

$$f_X(x) = \frac{1}{B(b-a)^{t-1}} (x-a)^{r-1}(b-x)^{t-r-1} \qquad a \leqslant X \leqslant b$$

where:

> a and b = the minimum and maximum values of X respectively.
> r and t = numerical constants related to the mean and variance of X.
> B = the normalising constant.

The beta distribution is usually symbolised as $BT(r, t)$ and the range of shapes of distribution curves that it can represent is remarkable, varying from rectangles to symmetrical or asymmetrical curves.

NECESSARY INFORMATION

The mathematics behind this distribution will not be examined but if the reader wishes to use the distribution he will require the following information.

$$m_X = a + \frac{r}{t}(b - a); \qquad \sigma_X{}^2 = \frac{(b - a)^2 r(t - r)}{t^2(r + 1)}$$

Provided that r and t are integers B can be found from the formula:

$$B = \frac{(r - 1)!\,(t - r - 1)!}{(t - 1)!}$$

if r and t are not whole numbers, approximation can be attempted but it is probably best to determine B by using the fact that the total area under the pdf curve is equal to one.

The procedure is to determine a suitable number of values of $f_X(x)$ over the range of X values using the formula, with B put equal to one.

With these values and by Simpson's rule, the area under the curve (with B equal to one) can be obtained. This value of the area will be the true value of B.

The method is simple to understand but involves a lot of tedious work which is best computerised.

VALUES OF a AND b

The minimum and maximum values of X, a and b, must be known if $f_X(x)$ is to be determined and suggested approximations for soil parameters are given later in this chapter.

Example 6.2

A series of measurements of the angle of shearing resistance of a sand determined that the parameter had a mean value of 35°, a standard deviation of 3° and that its minimum and maximum values were 28° and 42° respectively. Fit the results to a symmetrical beta distribution.

Solution

$$m_\phi = 35°; \qquad \sigma_\phi = 3°$$

Therefore:

$$35 = 28 + \frac{r}{t}(42 - 28) = 28 + 14\frac{r}{t}$$

It follows:

$$r = 0.5t \qquad\qquad (A)$$

so:

$$\sigma_\phi{}^2 = 3^2 = 14^2 \times \frac{0.5(t - 0.5)}{t^2(t - 1)}$$

Hence, t is 4.44 and, from equation (A), r is 2.22.
    If we assume t is 4 and r is 2 then:

$$B = \frac{(2 - 1)!\,(4 - 2 - 1)!}{(4 - 1)!} = \frac{1}{3!} = 0.1667$$

This figure would have to be used unless there was access to a microcomputer when a more accurate figure could be obtained by using the method suggested above. If this is done then B equals 0.1155.
    Using B equals 0.1155, values for $f_X(x)$ can be obtained for a suitable range of X values and are shown plotted in Fig. 6.1.

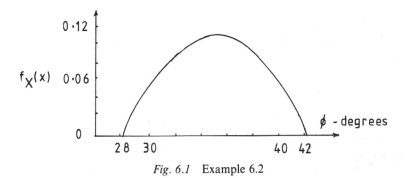

Fig. 6.1   Example 6.2

Example 6.3

Assuming that the mean and standard deviation values of example 3.10 are values of a saturated clay's unit cohesion, c_u, and that the minimum and maximum values of c_u are 22 and 55 kN/m² respectively, fit the results to a beta distribution.

Solution

a is 22 and b is 55, therefore the distribution is asymmetrical with the lower tail longer than the upper, i.e. it has a negative skew.

$$m_X = 44.85 \text{ kN/m}^2; \qquad \sigma_X = 6.56 \text{ kN/m}^2$$

Therefore:

$$44.85 = 22 + \frac{r}{t}(55 - 22) = 22 + 33\frac{r}{t} \tag{A}$$

$$\sigma_X{}^2 = 6.56^2 = (55 - 22)^2 \frac{r(t - r)}{t^2(r + 1)}$$

Hence, t is 4.39 and, from (A), r is 3.039.

For r = 3 and t = 4, $B = \dfrac{(3 - 1)!\,(4 - 3 - 1)!}{(4 - 1)!} = 0.3333$

For r = 3 and t = 5, $B = \dfrac{(3 - 1)!\,(5 - 3 - 1)!}{(5 - 1)!} = 0.0833$

Approximate value for B equals:

$$0.3333 - 0.39(0.3333 - 0.0833) = 0.2358$$

The suggested method gives B equal to 0.1850 and, with this value, the pdf is shown in Fig. 6.2.

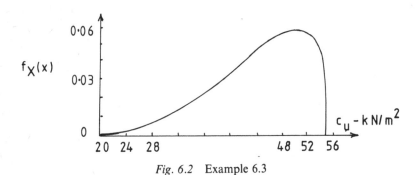

Fig. 6.2   Example 6.3

Estimation of soil distributions

Various workers have been involved in the study of the forms of distribution for different soil parameters together with the determination of the average value that their coeffcients of variation take. The main workers in this field have been Lumb (1966 and 1970), Schultze (1972), Turnbull et al. (1966), Meyerhof (1970), Hooper and Butler (1966) and Alonson (1976).

At the present time, for both undisturbed and compacted soils, the consensus of opinion is broadly as follows.

Many soil parameters have distributions that approximate to the normal, or lognormal, distribution. These parameters include:

- Density (dry and bulk)
- Voids ratio
- Water content
- Degree of saturation
- Liquid limit
- Plastic limit
- Plasticity index
- Particle specific gravity
- Coefficient of consolidation
- Angle of shear resistance

Parameters whose distributions do not often approximate very well to normal distributions include:

- Compression index
- Coefficient of volume decrease
- Permeability
- Cohesion

These notes will only consider soil strengths and the reader is referred to the listed references for information on the other parameters.

Distributions of soil strength parameters

TWO COMPONENT SOILS

The strength of unsaturated soils, such as silts, sandy clays, etc., are made up from two components:

(i) the unit cohesion, c
(ii) the tangent of the angle of friction, $\tan \phi$

For such soils the drained strength parameters give the best measure of strength and it should be noted that the symbols c and ϕ used here refer to the drained, or effective, cohesion and angle of friction of the soil.

(*i*) *Cohesion* This parameter best fits a positively skewed beta distribution. (Positively skewed means that the upper tail of the distribution is longer than the lower tail.)

It is apparent from the formula for $f_X(x)$ for the beta distribution that the minimum and maximum values, a and b, of the distribution must be known. Provided that these values have been obtained by test measurements there will be no problem but, bearing in mind that there will need to be about 30 measurements in order to obtain values of a and b to an acceptable level of significance, some form of estimation will generally be necessary.

For the asymmetrical beta distribution of the cohesion of a two component soil Lumb (1970) suggests the following approximate values for a and b:

$$a = m_c - 1.9\sigma_c; \qquad b = m_c + 4.1\sigma_c$$

(*ii*) *Friction* The distribution of the tangent of the effective angle of friction is more or less symmetrical and is close to a symmetrical beta distribution.

If no measured values of the minimum and maximum values of $\tan \phi$ are available, Lumb suggests the following approximations:

$$a = m_{\tan\phi} - 2.3\sigma_{\tan\phi}; \qquad b = m_{\tan\phi} + 2.3\sigma_{\tan\phi}$$

ONE COMPONENT SOILS

(*i*) *Cohesion* For a saturated clay the relevant strength component is the undrained cohesion, c_u, which is completely different to the drained cohesion and, as one would intuitively expect, it has an entirely different distribution.

The best approximation to this distribution is a negatively skewed beta distribution, i.e. an asymmetrical distribution with a longer tail of lower values.

Approximate values for the minimum and maximum values are:

$$a = m_{cu} - 3.5\sigma_{cu}; \qquad b = m_{cu} + 1.6\sigma_{cu}$$

(*ii*) *Friction* It has been found that the assumption of a symmetrical beta distribution, as for a two component soil, or of a normal distribution both give satisfactory approximations for the distribution of $\tan \phi$, where ϕ is the drained, or effective, angle of shearing resistance, although the beta distribution gives a closer fit for the lower tail.

Note: It has been shown, Lumb (1965) that variations in the effective angle of friction, ϕ, are statistically independent of grading parameters, void ratio and degree of saturation and that both ϕ or $\tan\phi$ can generally be taken as being normally distributed variables.

Variability of civil engineering parameters

DIMENSIONS

Variations in the main dimensions of a structure are usually of small magnitude and of less significance than variations in the strength or loading parameters to which the structure is subjected. For an average structure the coefficient of variation of the dimensions is considerably less than 5% and can often be ignored, i.e. structural dimensions are assumed to be constant.

These remarks apply to main dimensions such as the span and depth of a beam, the size of a footing etc. but it must be remembered that if there is a possibillity of a variation in a critical dimension, such as the depth of a reinforcement layer in a concrete beam, then this should be allowed for.

MANUFACTURED CONSTRUCTIONAL MATERIALS

With reasonable quality control the unit weights, strengths, etc. of manufactured materials have only small coefficients of variation, usually not more than 1%. As with dimensions, it can often be assumed that unit weights of materials such as steel, brick, or concrete are constant.

SOILS

Typical values for the coefficients of variation of the main soil parameters are listed below.

	Sand	Silt	Clay
Density			
(undisturbed soil)	5–10	5–10	5–10%
(compacted soil)	2.5–5	2.5–5	2.5–5%
Water content			
(undisturbed soil)	5	10–23	12–22%
(compacted soil)	5	5–12	6–11%
Void ratio	13–30	22	15–32%
Liquid limit		5.5	22–28%
Plastic limit		12	20–45%

Strength – one component soil

Soft clay ($c_u < 40\,\text{kN/m}^2$): undrained cohesion $V = 20 - 25\%$
Hard clay ($c_u > 40\,\text{kN/m}^2$): undrained cohesion $V = 20 - 35\%$
Sand: drained angle of shearing resistance $V = 5 - 15\%$

Strength – two component soil

Two component soils are extremely variable in strength. Typical coefficients of variation are given by Lumb (1974):

Clay shale: cohesion 95%; $\tan \phi = 46\%$
Cohesive till: cohesion 100%; $\tan \phi = 18\%$
Residual sands and silts: cohesion 17%; $\tan \phi = 6\%$

Time dependency of variables

The parameters considered so far have been single variables with values assumed to be independent of time. This assumption is usually justifiable in not only the reliability analyses of steel or concrete structures but also for soil structures.

For example, although the shear strength of a soil changes from undrained to drained, a reliability analysis is usually carried out using either the undrained or the drained strength values, depending on the design requirements.

For all practical purposes, the dimensions of a structure do not vary with time and the strengths and densities of constructional materials also can usually be assumed as non-time dependent, any variations being considered as a random process of the type discussed in earlier chapters. However, the possibility of finishing layers being progressively placed on top of existing layers leading to the phenomenon of a gradually increasing dead load should not be ignored.

Probability distributions of superimposed loadings

Superimposed loadings are obviously time dependent and vary considerably in their nature although there can be cases of predictable floor loadings where the assumption of a non-time dependent normal or lognormal distribution is realistic and will lead to satisfactory results.

Outside the above exceptions the effect of time on superimposed loading should be considered.

A superimposed load may involve a sudden change followed by a period in which the magnitude of the loading barely alters, such as the change in furniture and equipment loadings caused by the requirements of a new tenant in a rented industrial or office building. Such loadings have been discussed by Tang (1981), and their consideration is necessary in both the evaluation of long term effects such as settlement and in the study of how the structure is likely to behave when subjected to some form of extreme loading.

Extreme values of loading can occur within the continual and irregular series of load values set up by a natural agency, such as snow, wind, waves and earthquakes.

For design the main concern is about the largest, or extreme, load values that the structure is likely to be subjected to together with the minimum resistance that the structure will offer. This aspect of reliability analysis is considered in the following section.

THEORY OF EXTREME VALUES

For such loadings we are obliged to consider the statistical theory of extreme values.

As already mentioned extreme load values can arise from several different and independent causes, probably the most important being earthquakes, wind, waves and snow.

The formation of an extreme-value distribution can be illustrated numerically, as in the following example which deals with the possible variation of the value of wind velocity.

Example 6.4

At a certain geographical location the maximum weekly wind speed was noted for a period of a year and was found to be normally distributed with a mean of 65 km/hr and standard deviation of 16 km/hr.

Prepare a histogram of the maximum annual wind speed for a period of 50 years.

Solution
If, of course, there was a 50 year set of weekly readings then the maximum value for each year could be obtained and a histogram plotted. Not having such readings, but knowing that the weekly values are normally distributed, an approximation to the histogram can be obtained by Monte Carlo simulation.

The minimum and maximum wind speeds possible will be assumed to be the mean value ± 4 multiplied by the standard deviation. If this range is divided into cells of equal width, say 0.25, then the lowest cell will have the range from -3.75 to -4.00 and highest cell from 3.75 to 4.00.

The centre of the lowest cell is equal to -3.875 and, from appendix III, the probability corresponding to this value of z is 0.0001 so that if the computer is programmed to produce 10 000 random numbers between one and 10 000, then the probability 0.0001 equals the chance of it generating the number one. If the number one is produced then we assume a wind speed corresponding to the centre of the lowest cell, i.e. 65 $- 3.875 \times 16 = 3$ km/hr.

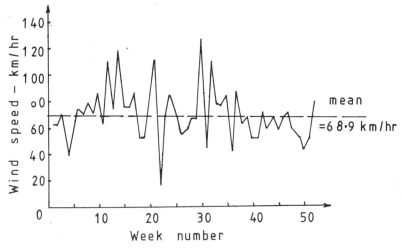

Fig. 6.3   Example 6.4 – simulated wind speeds for one year

Similarly if the computer produces the number 10 000 we assume a wind speed equal to $60 + 3.875 \times 16 = 127$ km/hr.

Within these two extremes a whole range of probable wind speeds can be generated. If, for example the number generated lies between 1056 and 1587, then it lies between the probabilities of 0.1056 and 0.1587, corresponding to z values of -1.25 and -1.00 and a number generated within this range is taken to represent a wind speed of $65 - 1.125 \times 16 = 47$ km/hr.

The program was prepared to produce 50 independent distributions of 52 independent values, to represent the 52 weeks of each of the 50 years, and a complete set of weekly readings for one year is set out below and illustrated in Fig. 6.3. The program could easily have been adjusted to produce 50 sets of 365 values, to represent daily readings but it was felt that such a large number of values would only confuse the example.

Week	Wind speed (km/hr)	Week	Wind speed (km/hr)	Week	Wind speed (km/hr)
1	63.0	2	63.0	3	71.0
4	39.0	5	59.0	6	75.0
7	71.0	8	79.0	9	71.0
10	87.0	11	63.0	12	111.0
13	75.0	14	119.0	15	75.0
16	75.0	17	87.0	18	51.0
19	51.0	20	71.0	21	111.0
22	15.0	23	71.0	24	87.0
25	71.0	26	55.0	27	59.0
28	67.0	29	67.0	30	127.0
31	43.0	32	111.0	33	67.0
34	67.0	35	75.0	36	39.0
37	87.0	38	63.0	39	67.0
40	51.0	41	51.0	42	71.0
43	59.0	44	67.0	45	59.0
46	67.0	47	71.0	48	55.0
49	51.0	50	43.0	51	51.0
52	79.0				

It is seen that, for this year, the maximum wind velocity was 127.0 km/hr.

The histogram is shown in Fig. 6.4 and demonstrates that the wind speed distribution is normal.

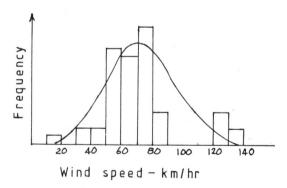

Fig. 6.4   Histogram of example 6.4

The procedure was now repeated another 49 times to yield the 50 values of maximum annual wind speed. Each of these values is a measurement of a new random variable Y, the variable that represents the extreme value of wind speed. These values are tabulated below.

Year	Max. speed (km/hr)	Year	Max speed (km/hr)	Year	Max. speed (km/hr)
1	87.0	2	103.0	3	115.0
4	95.0	5	95.0	6	99.0
7	111.0	8	99.0	9	103.0
10	99.0	11	99.0	12	99.0
13	103.0	14	91.0	15	91.0
16	95.0	17	103.0	18	91.0
19	111.0	20	107.0	21	107.0
22	99.0	23	127.0	24	103.0
25	107.0	26	91.0	27	111.0
28	107.0	29	103.0	30	99.0
31	99.0	32	91.0	33	99.0
34	107.0	35	95.0	36	111.0
37	99.0	38	103.0	39	119.0
40	99.0	41	103.0	42	127.0
43	95.0	44	99.0	45	107.0
46	103.0	47	107.0	48	91.0
49	99.0	50	111.0		

The histogram of these values of Y is shown in Fig. 6.5a. From this histogram it is possible to obtain an approximation to the pdf of this particular distribution and this is drawn in on the figure.

It is seen that although the parent distributions were all normal this distribution of extreme values is by no means normal. It is referred to as an extreme distribution.

By considering the pdf line drawn on Fig. 6.5a it is possible to slightly adjust the arrangement of the histogram cells so that they conform more closely with the sketched in pdf line. This has been done in Fig. 6.5b and it is now a simple matter to obtain an approximation for the $f_Y(y)$ and $F_Y(y)$ value bearing in mind that the total area of the histogram is equal to one.

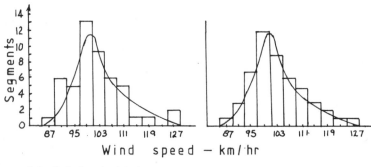

(a) Original histogram   (b) Modified histogram

Fig. 6.5   Example 6.4

There are 50 segments in the 11 cells. The first cell consists of one segment and the area of this cell is therefore $\frac{1}{50}$ (0.02).

Dividing the area of the cell by its width (4 km/hr) gives the central height of the cell, i.e. $f_Y(y)$ and is therefore 0.005. $F_Y(y)$ is obviously equal to 0.02.

The second cell has three segments so that $f_Y(y)$ equals $\frac{3}{50} \times 4 = 0.015$ and $F_Y(y)$ equals $(3 + 1)/50 = 0.08$.

The $f_Y(y)$ and $F_Y(y)$ values are set out below.

y (km/hr)	$f_Y(y)$	$F_Y(y)$
87	0.005	0.02
91	0.015	0.08
95	0.035	0.22
99	0.06	0.46
103	0.05	0.66
107	0.03	0.78
111	0.02	0.86
115	0.015	0.92
119	0.01	0.96
123	0.005	0.98
127	0.005	1.00

Extreme value distributions

Extreme distributions can obviously be obtained for either tail of the parent distribution, i.e. for maximum or minimum values. Probably the most valuable work carried out on extreme value statistics is that by Gumbel (1958).

The distribution from which the extreme values are generated is known as the parent distribution, an example being the yearly set of the weekly wind values of the preceeding example. The distribution of extremes can be easily expressed as a function of the parent distribution and the sample size, n, the number of times that the distribution was considered (i.e. 50 in example 6.1).

The probability that all of the n independent observations will be less than x is $[F_X(x)]^n$. This can be re-expressed as the probability that y, the largest value among the n independent observations, is less than or equal to x, $F_X(y)$. Hence:

$$F_Y(y) = [F_X(x)]^n$$

In other words the cumulative distribution function of the extreme value distribution is the cumulative distribution function of the parent distribution raised to the power of the sample size. Unfortunately the formula is of little practical significance owing to the high powers generally involved.

The significant property of the extreme distribution was discovered by Fisher and Tippet (1928), who showed that, as the sample size increases the actual form of the distribution asymptotically approaches one of three distinct forms, called type I, type II and type III.

Fisher and Tippet showed that the type of distribution that is finally achieved depends upon the properties of the tails of the parent distribution.

THE TYPE I DISTRIBUTION

If the parent distribution consists of a random variable, X, with a cumulative probability distribution of the form:

$$F_X(x) = 1 = \exp[-g(x)]$$

where $g(x)$ is a monotonic increasing function of x, then the extreme value distribution of Y is type I. (The normal and beta distributions have cdfs of this form.)

When considering the pdf of the type I distribution it is advantageous to think in terms of a reduced variable, $\alpha(y - u)$. u is the mode of the distribution and α is a measure of the dispersion.

It can be shown (Benjamin and Cornell 1970) that:

$$\alpha = \frac{1.282}{\sigma_Y}; \qquad u = m_Y - \frac{0.577}{\alpha}$$

The expressions for $f_Y(y)$ and $F_Y(y)$ then become:

$$f_Y(y) = \exp[-\alpha(y - u) - \exp(-\alpha(y - u))]$$
$$F_Y(y) = \exp[-\exp(-\alpha(y - u))]$$

For maximum values the distribution is positively skewed and has the form of the pdf estimated in Fig. 6.5b. For minimum values, the distribution is negatively skewed and can allow for values of y being less than zero.

It is generally agreed that the type I extreme distribution is the most suitable for civil engineering design work, which is usually concerned with the prediction of maximum values of snow and wind loading. It has also been used to model the long term distribution of North Sea wave heights (Saetre 1975).

However, for the sake of completeness, brief notes on the type II and the type III distributions are set out below.

TYPE II DISTRIBUTION

If the parent distribution has a cdf of the form:

$$F_X(x) = 1 - C\left(\frac{1}{x}\right)^k \qquad x \geqslant 0$$

then the extreme distribution is type II with the following expressions for $f_Y(y)$ and $F_Y(y)$:

$$f_Y(y) = \frac{k}{u}\left(\frac{u}{y}\right)^{k+1} \exp[-(u/y)^k] \qquad y \geqslant 0$$

$$F_Y(y) = \exp[-(u/y)^k] \qquad\qquad y \geqslant 0$$

A relationship exists between type I and type II distributions which is identical to the relationship between the normal and the lognormal distributions.

It can be shown that if Y has a type II extreme distribution then the variable Z, equal to $\ln(Y)$, has a type I extreme distribution.

TYPE III DISTRIBUTION

If the parent distribution has a fixed limit on its maximum value, x_{max} = w (say), so that the cdf of the parent distribution, in the region of w, has the form:

$$F_X(x) = 1 - c(w - x)^k \qquad x \leqslant w, k > 0$$

then the extreme distribution is type III with the following expression for $f_Y(y)$ and $F_Y(y)$:

$$f_Y(y) = \frac{k}{w - u}\left(\frac{w - y}{w - u}\right)^{k-1} \exp\left[-\left(\frac{w - y}{w - u}\right)^k\right] \qquad y \leqslant w$$

$$F_Y(y) = \exp\left[-\left(\frac{w - y}{w - u}\right)^k\right] \qquad y \leqslant w$$

Example 6.5

Measurements taken over several years at a geographical location indicate that the mean value and standard deviation of the maximum annual wind velocity are 102.3 and 8.5 km/hr respectively. Assuming a type I extreme distribution, plot the pdf and the cdf of the wind speed's maximum values.

Solution

The values of the mean and standard deviation are those of the set of simulated maximum values obtained in example 6.4.

$$\alpha = \frac{1.282}{\sigma_Y} = \frac{1.282}{8.5} = 0.1508$$

$$u = m_Y - 0.577/\alpha = 102.3 - 0.577/0.1508 = 98.47 \text{ km/hr}$$

Substituting suitable values for Y into the expressions for $f_Y(y)$ and $F_Y(y)$ leads to the following results:

y	$f_Y(y)$	$F_Y(y)$
87	0.003	0.003
91	0.021	0.041
95	0.047	0.177
99	0.055	0.393
103	0.046	0.604
107	0.032	0.762
111	0.020	0.864
115	0.011	0.924
119	0.006	0.958
123	0.004	0.977
127	0.002	0.988

The cdf and pdf plots are shown in Fig. 6.6 along with the estimated plots obtained in example 6.4.

Earthquake loadings

Because of the different types of earthquake source, i.e. fault lines, aerial sources and point sources, the maximum lateral force to which an earth dam may be subjected because of earthquake action cannot be modelled satisfactorily by the above extreme distributions.

Analytical models whereby the maximum acceleration at a site can be estimated have been prepared by Cornell (1968) and Der Kiureghian and Ang (1977).

Soil induced loading

A major portion of the loading carried by a soil structure is generated by the soil itself. These induced forces are:

- Vertical forces (due to the weight of the soil).
- Lateral forces (thrusts due to active or passive earth pressure).

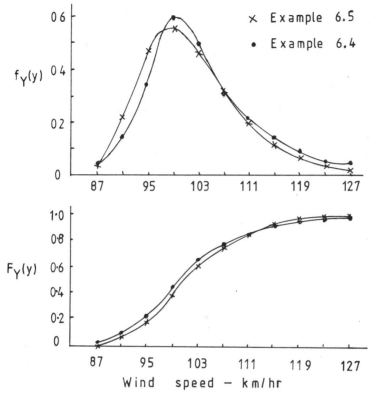

Fig. 6.6   pdf and cdf plots for example 6.4

It is seen from the examples in chapter 5 that the effects of such forces are included in the reliability analyses.

The one major soil variable that has not yet been mentioned is pore water pressure. In its simplest form the problem of pore water pressure is one of statics and concerns the hydrostatic state of soil below a ground water level. The pore water pressure in the soil is in a state of equilibrium and varies as the ground water level varies.

Ground water level variations are due to several agencies, such as seasonal temperature variations, rain and snow falls, constructional, irrigation and pumping schemes, etc. If records, such as piezometric readings, have been kept over a period it is possible to determine the maximum and minimum levels of ground water and to then fit them to some form of extreme distribution.

Such a situation is fairly unlikely and, for large and important structures, hydrological models must be used to predict ground water level changes (Uno et al. 1981).

For relatively small structures a ground water level is usually obtained from the site investigation report. Generally an estimation as to the highest ground water level can be made and this value is then taken as a constant in the reliability analysis.

For slopes, the coefficient of variation of the pore water pressure, and hence of r_u, is typically some 5 to 10% (Yuceman and Tang 1975).

Seepage of water through soil leads to seepage forces related to both the hydraulic gradient and the permeability of the soil. Any measured values of permeability can only be considered to be within plus or minus one order of magnitude and this can have a large effect on the accuracy of predictions from flow nets.

Mostyn (1983) has suggested that the use of the geometric, rather than arithmetic, mean of the permeability measurements can lead to more realistic flow diagrams. However, when one considers the inhomogenous nature of soil deposits, particularly silts and soils with the possibility of erratically positioned sand lenses, it is obvious that the prediction of seepage forces in a soil media is an intractable problem.

Chapter Seven

Matrix Algebra

In the final two chapters of this book it becomes apparent that a knowledge of matrices is essential for reliability analyses involving multivariates. This present chapter is a summary of those properties and mathematical operations of matrix algebra that are used in such analyses.

Matrices

A matrix is simply a collection of numbers, which may be real or complex. For example:

$$A = \begin{bmatrix} a & b & c \\ d & e & f \end{bmatrix} \quad \text{or} \quad B = \begin{bmatrix} 1 & 2 & 5 \\ 4 & 7 & 8 \\ 3 & 6 & 9 \end{bmatrix}$$

Generally a matrix is used in its entirety and, to save writing it out in full, it usually designated by a capital letter, A, B, C, etc.

The numbers of rows, m, in a matrix need not be equal to its number of columns, n. For instance both matrices A and B have three columns but A has two rows whereas B has three.

In general terms, a matrix that has m rows and n columns is said to be of the order $m \times n$ so that matrix A is said to be of the order 2×3 whereas matrix B is of the order 3×3.

Element of a matrix

Each term in a matrix is called an element. The convention used for the general element is a_{ij} which represents the element in both the ith row and the jth column. For example, in matrix B, a_{23} has the value 8.

Square matrix

A square matrix is one with its number of rows equal to its number of columns, i.e. $m = n$. Examples are:

$$A = \begin{bmatrix} 1 & 2 & 3 \\ 4 & 5 & 6 \\ 7 & 8 & 9 \end{bmatrix} \quad \text{or} \quad B = \begin{bmatrix} a_{11} & a_{12} & a_{13} \\ a_{21} & a_{22} & a_{23} \\ a_{31} & a_{32} & a_{33} \end{bmatrix}$$

Both these matrices are 3×3 matrices.

Leading diagonal

In a matrix the elements a_{ii}, for all values of i, are known as its leading elements.

In a square matrix the line from the top left-hand corner to the bottom right-hand corner of the matrix passes through all the leading elements, such as $a_{11}, a_{22}, a_{33}, a_{44}$ in the matrix below, and is referred to as the leading diagonal.

$$\begin{bmatrix} a_{11} & a_{12} & a_{13} & a_{14} \\ a_{21} & a_{22} & a_{23} & a_{24} \\ a_{31} & a_{32} & a_{33} & a_{34} \\ a_{41} & a_{42} & a_{43} & a_{44} \end{bmatrix}$$

Trace of a matrix

The sum of the elements of the leading diagonal is known as the trace of the matrix.

Diagonal matrix

A square matrix in which all the elements are zero except those on the leading diagonal is a diagonal matrix. For example:

$$\begin{bmatrix} 4 & 0 & 0 & 0 \\ 0 & 3 & 0 & 0 \\ 0 & 0 & 2 & 0 \\ 0 & 0 & 0 & 1 \end{bmatrix}$$

Unit matrix

A diagonal matrix in which all the non-zero elements are equal to one is a unit matrix. Example:

$$\begin{bmatrix} 1 & 0 & 0 & 0 \\ 0 & 1 & 0 & 0 \\ 0 & 0 & 1 & 0 \\ 0 & 0 & 0 & 1 \end{bmatrix}$$

Usually a unit matrix is given the symbol I.

Triangular matrix

An upper triangular matrix is a square matrix in which all the elements below the leading diagonal are zero. Example:

$$\begin{bmatrix} 2 & 6 & 7 & 9 \\ 0 & 3 & 6 & 1 \\ 0 & 0 & 1 & 4 \\ 0 & 0 & 0 & 3 \end{bmatrix}$$

A lower triangular matrix is a square matrix in which all the elements above the leading diagonal are zero.

Column matrix

A matrix which consists of a single column of values is known as a column matrix or as a column vector. Because such a matrix can represent the components of a vector it is usually designated by a small letter, a, b, c, etc. For example:

$$a = \begin{bmatrix} 14.6 \\ 28.7 \\ 12.5 \\ 2.9 \\ 13.4 \end{bmatrix}$$

Addition and subtraction of matrices

These are simple operations but are only possible if the two matrices, or vectors, involved, are of the same order. The corresponding elements are added (or subtracted) to give the corresponding element in the resultant matrix. For example:

$$\begin{bmatrix} 2 & -5 & 6 \\ 4 & 7 & 2 \end{bmatrix} + \begin{bmatrix} 3 & 8 & 1 \\ 9 & -3 & 4 \end{bmatrix} = \begin{bmatrix} 5 & 3 & 7 \\ 13 & 4 & 6 \end{bmatrix}$$

$$\begin{bmatrix} 2 & -5 & 6 \\ 4 & 7 & 2 \end{bmatrix} - \begin{bmatrix} 3 & 8 & 1 \\ 9 & -3 & 4 \end{bmatrix} = \begin{bmatrix} -1 & -13 & 5 \\ -5 & 10 & -2 \end{bmatrix}$$

Multiplication of a matrix by a scalar quantity

The matrix $B = kA$ is obtained by simply multiplying every element of A by k. For example:

$$3 \begin{bmatrix} 3 & 6 & 2 \\ -12 & 4 & -3 \\ 10 & 1 & 8 \end{bmatrix} = \begin{bmatrix} 9 & 18 & 6 \\ -36 & 12 & -9 \\ 30 & 3 & 24 \end{bmatrix}$$

Multiplication of matrices

Two matrices, A and B, may only be multiplied together if the number of columns of A equals the number of rows of B.

If A multiplies B then c_{ij}, the general element of the product matrix C, is obtained by multiplying the jth column of B by the ith row of A. Hence, if matrix A is of order $m \times p$ and matrix B of order $p \times n$ then the product matrix C is of order $m \times n$.

Example 7.1

Determine C given:

$$C = \begin{bmatrix} 3 & -2 & 1 & 4 \\ 4 & 12 & 6 & 3 \end{bmatrix} \begin{bmatrix} 3 & 5 & 6 \\ 1 & -4 & 3 \\ 6 & 2 & 5 \\ 8 & -9 & 5 \end{bmatrix}$$

Solution

Consider the first row of matrix C:

$$c_{11}, c_{12}, c_{13}$$

To determine c_{11}, the first row of A multiplies the first column of B:

$$c_{11} = 3 \times 3 + (-2) \times 1 + 1 \times 6 + 4 \times 8 = 45$$

To determine c_{12}, the first row of A multiples the second column of B:

$$c_{12} = 3 \times 5 + (-2) \times (-4) + 1 \times 2 + 4 \times (-9) = -11$$

Similarly:

$$c_{13} = 3 \times 6 + (-2) \times 3 + 1 \times 5 + 4 \times 5 = 37$$

To determine the second row of C, i.e. c_{21}, c_{22}, c_{23}, we multiply the first, second and third columns of B by the second row of A:

$$c_{21} = 4 \times 3 + 12 \times 1 + 6 \times 6 + 3 \times 8 = 84$$

$$c_{22} = 4 \times 5 + 12 \times (-4) + 6 \times 2 + 3 \times (-9) = -43$$

$$c_{23} = 4 \times 6 + 12 \times 3 + 6 \times 5 + 3 \times 5 = 105$$

Hence:

$$C = \begin{bmatrix} 45 & -11 & 37 \\ 84 & -43 & 105 \end{bmatrix}$$

FURTHER NOTES ON MATRIX MULTIPLICATION

It may have been noticed that the matrix operations of example 7.1 were carried out working from left to right and we say that matrix A premultiplies B or, alternatively, matrix B postmultiplies matrix A.

In matrix multiplication the matrices involved are distributive and associative but not commutative. This statement is probably best explained by considering three matrices A, B and C multiplied together in various ways:

$$ABC = (AB)C = A(BC)$$

$$A(B + C) = AB + AC$$

But:

$$ABC \neq CAB$$

It is seen from the above that matrix multiplication must always be carried out from left to right as the product matrix AB is not necessarily equal to BA indeed, in the case of example 7.1, it is just not possible to multiply matrix A by matrix B.

Transpose of a matrix

Associated with any matrix A is another matrix A^T known as its transpose. The transpose matrix is formed by changing the rows into columns and the columns into rows. For example:

$$A = \begin{bmatrix} 5 & 4 & 7 & 8 \\ 6 & 7 & -4 & 2 \\ 1 & -4 & 9 & 4 \\ 6 & -6 & 5 & -3 \end{bmatrix} \quad A^T = \begin{bmatrix} 5 & 6 & 1 & 6 \\ 4 & 7 & -4 & -6 \\ 7 & -4 & 9 & 5 \\ 8 & 2 & 4 & -3 \end{bmatrix}$$

By selecting suitable matrices the reader can verify a useful law that is associated with transposes:

$$(AB)^T = B^T A^T$$

A more general form of this law is:

$$(A_1 A_2 A_3 \ldots A_n)^T = A_n{}^T A_{n-1}{}^T \ldots A_3{}^T A_2{}^T A_1{}^T$$

Symmetric matrix

A matrix that is equal to its transpose is known as a symmetric matrix. For example:

$$A = \begin{bmatrix} 2 & 3 & -4 \\ 3 & 5 & 6 \\ -4 & 6 & 7 \end{bmatrix} \quad A^T = \begin{bmatrix} 2 & 3 & -4 \\ 3 & 5 & 6 \\ -4 & 6 & 7 \end{bmatrix}$$

It should be noted that the product of two symmetric matrices is not necessarily a symmetric matrix.

Null matrix

If all the elements of a matrix are equal to zero then the matrix is referred to as a null matrix. Similarly a vector consisting of zero value elements is called a null vector.

Multiplication of vectors

Two vectors, x and y, when written in their normal column vector form, cannot be multiplied by the rules we have established. For example:

$$\begin{bmatrix} x_1 \\ x_2 \\ x_3 \\ \vdots \\ x_n \end{bmatrix} \begin{bmatrix} y_1 \\ y_2 \\ y_3 \\ \vdots \\ y_n \end{bmatrix}$$

cannot be multiplied together. It is necessary to write the vectors in the form:

$$\begin{bmatrix} x_1 & x_2 & x_3 & \cdots & x_n \end{bmatrix} \begin{bmatrix} y_1 \\ y_2 \\ y_3 \\ \vdots \\ y_n \end{bmatrix}$$

Multiplication is now possible, provided that the number of columns in the left-hand vector equals the number of rows in the right-hand one.

Hence, in order to establish the product xy we must use $x^T y$. Using the transpose of x does not alter the vector in any way and is simply carried out in order to make the multiplication possible.

It is obvious that $x^T y = y^T x$. This is not the case with matrices and the rule can only be applied to vectors.

The determinant of a square matrix

With every square matrix there is an arithmetical number, called the determinant of the matrix, associated with it. If a matrix A is of order $n \times n$ then its determinant, $|A|$, can be obtained from the formula:

$$|A| = a_{11}M_{11} - a_{12}M_{12} + a_{13}M_{13} + \cdots + (-1)^n a_{1n}M_{1n}$$

where M_{11} is the minor of a_{11}, i.e. the determinant of the matrix when row 1 and column 1 are removed.

M_{12} is the minor of a_{12}, i.e. the determinant of the matrix when row 1 and column 2 are removed.

M_{1n} is the minor of a_{1n}, i.e. the determinant of the matrix when row 1 and column n are removed.

In order to use the formula it must be remembered that a single element, such as a_{11}, is really a 1×1 matrix and that its determinant is equal to itself, i.e. a_{11}.

The formula becomes even simpler if it is expressed in terms of cofactors. The cofactor of a_{ij}, C_{ij}, is simply the minor of a_{ij} multiplied by $(-1)^{i+j}$. Then:

$$|A| = a_{11}C_{11} + a_{12}C_{12} + a_{13}C_{13} + \cdots + a_{1n}C_{1n}$$

Eample 7.2

Determine |A| given:

$$A = \begin{bmatrix} a_{11} & a_{12} \\ a_{21} & a_{22} \end{bmatrix}$$

Now:

$$C_{11} = (-1)^{1+1}(\text{Minor } a_{11}) = (-1)^2|a_{22}| = a_{21}$$

and:

$$C_{12} = (-1)^{1+2}(\text{Minor } a_{12}) = (-1)^3|a_{11}| = -a_{21}$$

Hence:

$$|A| = a_{11}a_{22} - a_{12}a_{11}$$

Example 7.3

Determine |A| given:

$$A = \begin{bmatrix} 5 & 8 & 10 \\ 0 & -2 & 6 \\ 3 & 12 & 14 \end{bmatrix}$$

The formula quoted for |A| can apply to any row and indeed to any column. As the first column of A includes a zero term it is simplest to use the formula for the first column thus:

$$|A| = a_{11}C_{11} + a_{21}C_{21} + a_{31}C_{31}$$

$$C_{11} = [-2 \times 14 - 6 \times 12][-1]^{(1+1)} = -100$$

$$C_{31} = [8 \times 6 - 10 \times (-2)][-1]^{(1+3)} = 68$$

Hence:

$$|A| = -5 \times 100 + 3 \times 68 = -296$$

EVALUATION OF A DETERMINANT BY ELIMINATION

Obviously the amount of computation with the cofactor method rapidly increases with the size of the matrix and for large matrices the Gaussian method of elimination, which can be computerised, is useful. The method is based on the principle that when a particular row (or column) of a determinant has a scalar multiple of another row (or column) either added to it or subtracted from it, the value of the determinant is unaltered. If this procedure is continually applied, an upper triangular matrix is eventually produced and the determinant is simply equal to the products of the elements on the leading diagonal, all other terms in the determinant involving a zero term.

Example 7.4

Evaluate

$$|A| = \begin{vmatrix} 3 & 1 & 6 & 3 \\ 5 & 1 & 3 & 2 \\ 4 & 2 & 0 & 2 \\ 3 & 5 & 9 & 1 \end{vmatrix}$$

Solution

Put row(2) = row(2) $- \frac{5}{3} \times$ row(1)
Put row(3) = row(3) $- \frac{4}{3} \times$ row(1)
Put row(4) = row(4) $-$ row(1)

$$\begin{vmatrix} 3 & 1 & 6 & 3 \\ 0 & -\frac{2}{3} & -7 & -3 \\ 0 & \frac{2}{3} & -8 & -2 \\ 0 & 4 & 3 & -2 \end{vmatrix}$$

Put row(3) = row(3) + row(2)
Put row(4) = row(4) + 6 $\times$ row(2)

$$\begin{vmatrix} 3 & 1 & 6 & 3 \\ 0 & -\frac{2}{3} & -7 & -3 \\ 0 & 0 & -15 & -5 \\ 0 & 0 & -39 & -20 \end{vmatrix}$$

Put row(4) = row(4) − $\frac{39}{15}$ × row(3)

$$\begin{vmatrix} 3 & 1 & 6 & 3 \\ 0 & -\frac{2}{3} & -7 & -3 \\ 0 & 0 & -15 & -5 \\ 0 & 0 & 0 & -7 \end{vmatrix}$$

Hence:

$$|A| = 3 \times (-\tfrac{2}{3}) \times (-15) \times (-7) = -210$$

The inverse matrix

Simultaneous equations are often encountered in civil engineering and, as in the examples illustrated in this book, are often of the first degree.

Let us consider a set of first degree equations that relate two vectors x and y as follows:

$$y_1 = a_{11}x_1 + a_{12}x_2 + a_{13}x_3 + \cdots + a_{1n}x_n$$

$$y_2 = a_{21}x_1 + a_{22}x_2 + a_{23}x_3 + \cdots + a_{2n}x_n$$

$$y_3 = a_{31}x_1 + a_{32}x_2 + a_{33}x_3 + \cdots + a_{3n}x_n$$

$$\cdot \qquad \cdot \qquad \cdot \qquad \cdot \qquad \cdots \qquad \cdot$$

$$\cdot \qquad \cdot \qquad \cdot \qquad \cdot \qquad \cdots \qquad \cdot$$

$$y_n = a_{n1}x_1 + a_{n2}x_2 + a_{n3}x_3 + \cdots + a_{nn}x_n$$

These relationships can be expressed in matrix form as:

$$y = Ax$$

where:

$$y = \begin{bmatrix} y_1 \\ y_2 \\ y_3 \\ \vdots \\ y_n \end{bmatrix} \qquad A = \begin{bmatrix} a_{11} & a_{12} & a_{13} & \cdots & a_{1n} \\ a_{21} & a_{22} & a_{23} & \cdots & a_{2n} \\ a_{31} & a_{32} & a_{33} & \cdots & a_{3n} \\ \vdots & \vdots & \vdots & \cdots & \vdots \\ a_{n1} & a_{n2} & a_{n3} & \cdots & a_{nn} \end{bmatrix} \qquad x = \begin{bmatrix} x_1 \\ x_2 \\ x_3 \\ \vdots \\ x_n \end{bmatrix}$$

If in the above simultaneous equations y is not a null vector then they are known as non-homogeneous equations and there must be an inverse relationship relating x to y.

In matrix terms this relationship is:

$$x = A^{-1}y$$

A^{-1} is known as the inverse matrix of A.

We therefore have two expressions:

$$y = Ax \tag{1}$$

$$x = A^{-1}y \tag{2}$$

Substituting for y in equation (2):

$$x = AA^{-1}x$$

which means that the product AA^{-1} must be a matrix which, when it multiplies x, it does not alter it in any way.

The only matrix with such a property is the unit matrix, I, which can either premultiply or postmultiply another matrix or vector without altering any values. Hence:

$$AA^{-1} = A^{-1}A = I$$

A matrix when multiplied by its inverse yields a unit matrix. This can often be a useful check during the evaluation of an inverse.

Singular matrix

It is always possible that A^{-1} does not exist, in other words, although it is possible to transform the vector x into the vector y, it is not possible to transform the vector y into the vector x. In such instances the matrix A is said to be singular.

A matrix is singular if it is rectangular or if its determinant is equal to zero.

Evaluation of the inverse matrix

In order to evaluate the inverse of a matrix, A, it is necessary to understand two other matrix forms of A.

COFACTOR MATRIX

This matrix is generally given the symbol A^c and is formed by replacing every element of the original matrix by its cofactor.

Example 7,5

If:

$$A = \begin{bmatrix} 6 & 2 & 5 \\ 3 & 8 & 2 \\ 5 & 4 & 7 \end{bmatrix}$$

then the cofactor matrix, A^c, can be established from the procedures already described and is:

$$A^c = \begin{bmatrix} 48 & -11 & -28 \\ 6 & 17 & -14 \\ -36 & 3 & 42 \end{bmatrix}$$

THE ADJOINT MATRIX

The adjoint matrix, A^a, is simply the transpose of the cofactor matrix. For example 7.5 the adjoint matrix is:

$$A^a = \begin{bmatrix} 48 & 6 & -36 \\ -11 & 17 & 3 \\ -28 & -14 & 42 \end{bmatrix}$$

It can be shown that A^{-1} equals $A^a/|A|$ which indicates why A can have no inverse when $|A|$ equals zero.

Example 7.6

Determine the inverse of matrix A of example 7.5.

Solution
By the methods already described, $|A|$ equals 126.

$$A^{-1} = \frac{A^a}{|A|}$$

$$= \frac{1}{126} \begin{bmatrix} 48 & 6 & -36 \\ -11 & 17 & 3 \\ -28 & -14 & 42 \end{bmatrix}$$

$$= \begin{bmatrix} 0.381 & 0.048 & -0.286 \\ -0.087 & 0.135 & 0.024 \\ -0.222 & -0.111 & 0.333 \end{bmatrix}$$

Example 7.7

Determine the vector y given the matrix relationship:

$$
\begin{bmatrix} 5 \\ 6 \\ -2 \end{bmatrix} = \begin{bmatrix} 6 & 2 & 5 \\ 3 & 8 & 2 \\ 5 & 4 & 7 \end{bmatrix} \begin{bmatrix} y_1 \\ y_2 \\ y_3 \end{bmatrix}
$$

The inverse of the matrix has been established in example 7.6. Hence we can write:

$$
\begin{bmatrix} y_1 \\ y_2 \\ y_3 \end{bmatrix} = \begin{bmatrix} 0.381 & 0.048 & -0.286 \\ -0.087 & 0.135 & 0.024 \\ -0.222 & -0.111 & 0.333 \end{bmatrix} \begin{bmatrix} 5 \\ 6 \\ -2 \end{bmatrix} = \begin{bmatrix} 2.765 \\ 0.327 \\ -2.442 \end{bmatrix}
$$

DETERMINATION OF THE INVERSE MATRIX BY ELIMINATION

In a similar way to a determinant, the use of the cofactor matrix for the determination of the inverse of a matrix becomes cumbersome as the order of the matrix increases. In such cases an elimination approach can be used and the technique is illustrated in the following example.

Example 7.8

Using an elimination procedure determine the inverse of matrix A in example 7.5.

Solution
The procedure is started by writing down the matrix to be inverted and, alongside it, a unit matrix of the same order:

$$
\begin{bmatrix} 6 & 2 & 5 \\ 3 & 8 & 2 \\ 5 & 4 & 7 \end{bmatrix} \begin{bmatrix} 1 & 0 & 0 \\ 0 & 1 & 0 \\ 0 & 0 & 1 \end{bmatrix}
$$

The process of finding the inverse consists of gradually changing the left-hand matrix into a unit matrix. Every change made to the left-hand matrix must be repeated in the right-hand matrix. At the end of the operations the right-hand matrix will be the inverse of the original matrix.

There are various routes that can be taken to the final result and the point to remember is that you are working towards achieving zero values above and below the leading diagonal of the left-hand matrix.

Put row(2) = 2 × row(2) − row(1)

$$\begin{bmatrix} 6 & 2 & 5 \\ 0 & 14 & -1 \\ 5 & 4 & 7 \end{bmatrix} \begin{bmatrix} 1 & 0 & 0 \\ -1 & 2 & 0 \\ 0 & 0 & 1 \end{bmatrix}$$

Put row(3) = 6 × row(3) − 5 × row(1)

$$\begin{bmatrix} 6 & 2 & 5 \\ 0 & 14 & -1 \\ 0 & 14 & 17 \end{bmatrix} \begin{bmatrix} 1 & 0 & 0 \\ -1 & 2 & 0 \\ -5 & 0 & 6 \end{bmatrix}$$

Put row(3) = row(3) − row(2)

$$\begin{bmatrix} 6 & 2 & 5 \\ 0 & 14 & -1 \\ 0 & 0 & 18 \end{bmatrix} \begin{bmatrix} 1 & 0 & 0 \\ -1 & 2 & 0 \\ -4 & -2 & 6 \end{bmatrix}$$

Put row(1) = 7 × row(1) − row(2)

$$\begin{bmatrix} 42 & 0 & 36 \\ 0 & 14 & -1 \\ 0 & 0 & 18 \end{bmatrix} \begin{bmatrix} 8 & -2 & 0 \\ -1 & 2 & 0 \\ -4 & -2 & 6 \end{bmatrix}$$

Put row(1) = row(1) − 2 × row(4)
Put row(2) = 18 × row(2) + row(3)

$$\begin{bmatrix} 42 & 0 & 0 \\ 0 & 252 & 0 \\ 0 & 0 & 18 \end{bmatrix} \begin{bmatrix} 16 & 2 & -12 \\ -22 & 34 & 6 \\ -4 & -2 & 6 \end{bmatrix}$$

put row(1) = row(1)/42;   row(2) = row(2)/252;   row(3) = row(3)/18

$$\begin{bmatrix} 1 & 0 & 0 \\ 0 & 1 & 0 \\ 0 & 0 & 1 \end{bmatrix} \begin{bmatrix} 0.381 & 0.048 & -0.286 \\ -0.087 & 0.135 & 0.024 \\ -0.222 & -0.111 & 0.333 \end{bmatrix}$$

Eigenvalues and eigenvectors

In many engineering applications one encounters simultaneous equations that have the matrix form:

$$Ax = \lambda x$$

where:

A = a square matrix
x = a column vector
λ = a scalar quantity

There is often more than one value of λ that satisfies the equation and these values are known as the characteristic roots, or eigenvalues, of the matrix A.

For each different value of λ there is a different vector, x, that satisfies the simultaneous equations and these vectors are known as the eigenvectors of matrix A.

For the work associated with this book the matrix A will be symmetrical.

Example 7.9

In a particular situation the following relationship applies:

$$\begin{bmatrix} 25.0 & 6.25 \\ 6.25 & 25.0 \end{bmatrix} \begin{bmatrix} x_1 \\ x_2 \end{bmatrix} = \lambda \begin{bmatrix} x_1 \\ x_2 \end{bmatrix}$$

Determine the eigenvalues and eigenvectors of the matrix.

Solution
The equation is of the form:

$$Ax = \lambda x$$

where:

$$A = \begin{bmatrix} 25.0 & 6.25 \\ 6.25 & 25.0 \end{bmatrix} \quad \text{and} \quad x = \begin{bmatrix} x_1 \\ x_2 \end{bmatrix}$$

Now:

$$[A - I\lambda] = \begin{bmatrix} 25.0 & 6.25 \\ 6.25 & 25.0 \end{bmatrix} - \lambda \begin{bmatrix} 1 & 0 \\ 0 & 1 \end{bmatrix}$$

$$= \begin{bmatrix} 25 - \lambda & 6.25 \\ 6.25 & 25 - \lambda \end{bmatrix}$$

The two simultaneous equations are:

$$25x_1 + 6.25x_2 = \lambda x_1$$

$$6.25x_1 + 25x_2 = \lambda x_2$$

which can be expressed in matrix form as:

$$\begin{bmatrix} 25 - \lambda & 6.25 \\ 6.25 & 25 - \lambda \end{bmatrix} \begin{bmatrix} x_1 \\ x_2 \end{bmatrix} = 0$$

which is seen to be of the form $[A - I\lambda]x = 0$.

The equation $[A - I\lambda]x = 0$ is the same as $Ax = \lambda x$ and is known as the characteristic equation of the matrix.

The trivial solution for this equation, and one that is correct, is that all the elements of x equal zero, i.e. that x is a null vector. If this is the only solution the matrix $[A - I\lambda]$ must have an inverse as this then establishes that $x = 0$.

It is therefore seen that for there to be any non-trivial solutions, the matrix A cannot have an inverse. It must be singular and therefore its determinant must be equal to zero. Hence, we can write:

$$(25 - \lambda)(25 - \lambda) - 6.25 \times 6.25 = 0$$

Therefore:

$$\lambda^2 - 50\lambda + 585.94 = 0$$

and:

$$\lambda = 31.25 \quad \text{or} \quad \lambda = 18.75$$

Hence, the eigenvalues of matrix A are 31.25 and 18.75.

A numerical check on the values of the eigenvalues can be obtained from the fact that the sum of the eigenvalues is equal to the trace of the original matrix:

$$31.25 + 18.75 = 50$$

$$25 + 25 = 50$$

Having obtained the eigenvalues of the matrix it is now possible to obtain the corresponding eigenvalues by substituting each value of λ into the equation.

The eigenvector corresponding to $\lambda = 31.25$ is obtained from the equations:

$$\begin{bmatrix} 25 - 31.25 & 6.25 \\ 6.25 & 25 - 31.25 \end{bmatrix} \begin{bmatrix} x_1 \\ x_2 \end{bmatrix} = 0$$

which gives the single equation:

$$-6.25x_1 + 6.25x_2 = 0$$

i.e. x_1 equals x_2.

So the general solution for the eigenvector is:

$$\alpha_1 \begin{bmatrix} 1 \\ 1 \end{bmatrix}$$

where α_1 is an arbitrary constant.

Similarly for the eigenvector corresponding to $\lambda = 18.75$ we obtain the relationship:

$$x_1 = -x_2$$

and, if we assume that x_2 is equal to one, makes the eigenvector equal to:

$$\alpha_2 \begin{bmatrix} -1 \\ 1 \end{bmatrix}$$

(The reader could have assumed that x_1 equals one in which case x_2 would have been -1. Subsequent calculations are unaffected by the choice as the relationship between x_1 and x_2 is still fully defined.)

A square $n \times n$ matrix obviously has n eigenvectors and once they have been obtained are written as columns in a square matrix, known as the modal matrix and given the symbol M. (An eigenvector is sometimes referred to as a modal column.) In the example:

$$M = \begin{bmatrix} 1 & -1 \\ 1 & 1 \end{bmatrix}$$

(putting α_1 and α_2 equal to one).

A popular form of modal matrix is when the eigenvectors are normalised, i.e. each element is divided by the length of the eigenvector.

$$\text{length of both vectors} = \sqrt{1^2 + 1^2} = \sqrt{2}$$

Hence:

$$N = \begin{bmatrix} 0.7071 & -0.7071 \\ 0.7071 & 0.7071 \end{bmatrix}$$

Note the use of the symbol N instead of M for normalised eigenvectors.

Spectral matrix

This particular matrix is important in the treatment of correlated variables, as described in chapter 8. It is simply a diagonal matrix in which the elements of the leading diagonal are the eigenvalues of the original matrix. It is generally given the symbol S. In example 7.9:

$$S = \begin{bmatrix} 31.25 & 0.0 \\ 0.0 & 18.75 \end{bmatrix}$$

Eigenvalues by transformation methods

The characteristic equation method can be extended to a 3×3 matrix, which involves a cubic expression, but the method is not suitable for large matrices and a technique that diagonalises the matrix and can be computerised is really the best approach.

There are various procedures available and Jacobi's method will be described here. This technique can only be applied to symmetrical matrices and is therefore suitable for the matrices discussed in this book with the advantage that the amount of computation involved is reduced.

The method involves an iterative procedure whereby the matrix is transformed into a diagonal one by the gradual elimination of the off-diagonal elements. In theory the procedure can go on for ever and it is therefore terminated once the values have achieved some required accuracy. Once the matrix is in a diagonal form its eigenvalues are the elements on the diagonal.

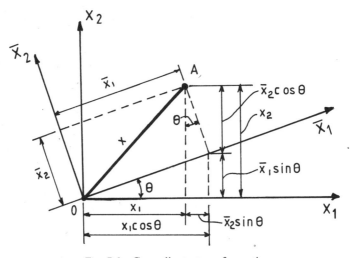

Fig. 7.1   Co-ordinate transformation

Consider a vector, OA, which in one set of coordinate axes has components x_1 and x_2 (Fig. 7.1). Let a further set of coordinate axes be at angle θ to Ox_1 and Ox_2 then the components of OA with reference to these axes are $\bar{x}_1$ and $\bar{x}_2$.

Considering Fig. 7.1 we see that:

$$x_1 = \bar{x}_1 \cos \theta - \bar{x}_2 \sin \theta$$

$$x_2 = \bar{x}_1 \sin \theta + \bar{x}_2 \cos \theta$$

i.e.

$$\begin{bmatrix} x_1 \\ x_2 \end{bmatrix} = \begin{bmatrix} \cos\theta & -\sin\theta \\ \sin\theta & \cos\theta \end{bmatrix} \begin{bmatrix} \bar{x}_1 \\ \bar{x}_2 \end{bmatrix}$$

which is a relationship of the form $x = R\bar{x}$.

Now, from the characteristic equation we know that $Ax = \lambda x$ so we can write:

$$AR\bar{x} = \lambda R\bar{x}$$

and multiplying both sides by R^T, the transpose of R, gives:

$$R^T AR\bar{x} = \lambda R^T R\bar{x}$$

Now:

$$R^T R = \begin{bmatrix} \cos\theta & \sin\theta \\ -\sin\theta & \cos\theta \end{bmatrix} \begin{bmatrix} \cos\theta & -\sin\theta \\ \sin\theta & \cos\theta \end{bmatrix} = \begin{bmatrix} 1 & 0 \\ 0 & 1 \end{bmatrix} = I$$

Hence:

$$R^T AR\bar{x} = \lambda I\bar{x} = \lambda\bar{x}$$

Let the transferred matrix $R^T AR$ be given the symbol B, then:

$$B = \begin{bmatrix} \cos\theta & \sin\theta \\ -\sin\theta & \cos\theta \end{bmatrix} \begin{bmatrix} a_{11} & a_{12} \\ a_{21} & a_{22} \end{bmatrix} \begin{bmatrix} \cos\theta & -\sin\theta \\ \sin\theta & \cos\theta \end{bmatrix} = \begin{bmatrix} b_{11} & b_{12} \\ b_{21} & b_{22} \end{bmatrix}$$

where:

$$b_{11} = a_{11}\cos^2\theta + 2a_{12}\sin\theta\cos\theta + a_{22}\sin^2\theta$$
$$b_{12} = a_{12}(\cos^2\theta - \sin^2\theta) + \sin\theta\cos\theta(a_{22} - a_{11})$$
$$b_{21} = a_{12}(\cos^2\theta - \sin^2\theta) + \sin\theta\cos\theta(a_{22} - a_{11})$$
$$b_{22} = a_{11}\sin^2\theta - 2a_{12}\sin\theta\cos\theta + a_{22}\cos^2\theta$$

It is possible to choose a value for θ so that the off-diagonal elements b_{12} and b_{21} become equal to zero. This value of θ can be obtained by equating b_{12}, or b_{21}, to zero.

$$a_{12}(\cos^2\theta - \sin^2\theta) + \sin\theta\cos\theta(a_{22} - a_{11}) = 0$$

Therefore:

$$\frac{a_{12}}{a_{11} - a_{22}} = \frac{\sin\theta\cos\theta}{\cos^2\theta - \sin^2\theta}$$

and so:

$$\frac{2a_{12}}{a_{11} - a_{22}} = \frac{2\sin\theta\cos\theta}{\cos^2\theta - \sin^2\theta} = \frac{\sin 2\theta}{\cos 2\theta}$$

Thus:

$$\tan 2\theta = \frac{2a_{12}}{a_{11} - a_{22}}$$

Using this value for θ, matrix B is:

$$\begin{bmatrix} b_{11} & 0 \\ 0 & b_{22} \end{bmatrix}$$

where b_{11} and b_{22} are the eigenvalues of matrix A.

Example 7.10

Determine the eigenvalues for matrix A of example 7.9.

Solution

$$A = \begin{bmatrix} 25.0 & 6.25 \\ 6.25 & 25.0 \end{bmatrix}$$

Therefore:

$$\tan 2\theta = \frac{2 \times 6.25}{25 - 25} = \infty$$

And so:

$$2\theta = 90°; \qquad \theta = 45°$$

Using these values:

$$b_{11} = 25\cos^2 45° + 2 \times 6.25\sin 45°\cos 45° + 25\sin^2 45° = 31.25$$

$$b_{22} = 25\sin^2 45° - 2 \times 6.25\sin 45°\cos 45° + 25\cos^2 45° = 18.75$$

which are the eigenvalues of matrix A.

Treatment of large order matrices

Example 7.10 deals with a two-dimensional vector, i.e. having only two components. For matrices of third order and above the various vectors will have three or more components. However, the principle of coordinate transformation can still be applied provided that the off-diagonal elements of the matrix are eliminated one at a time.

The final diagonalised matrix, B, is obtained after n iterations where n is the number of operations necessary to reduce all off-diagonal elements to zero. Hence:

$$B = T_n^T . T_{n-1}^T \cdots T_3^T . T_2^T . T_1^T [A] T_1 . T_2 . T_3 \cdots T_{n-1} . T_n$$

Example 7.11

Determine the eigenvalues of A:

$$A = \begin{bmatrix} 3 & 6 & 1 \\ 6 & 4 & 0 \\ 1 & 0 & 2 \end{bmatrix}$$

Solution
Using general terms:

$$\tan 2\theta = \frac{2a_{ij}}{a_{ii} - a_{jj}}$$

The first step is to select a suitable accuracy, say four decimal points.

First iteration
Select the off-diagonal element with the largest absolute value, in this case a_{12} equal to 6. Then:

$$\tan 2\theta = \frac{2a_{12}}{a_{11} - a_{22}} = \frac{2 \times 6}{3 - 4} = -12$$

Therefore:

$$2\theta = -85.24°; \quad \theta = -42.62°; \quad \sin \theta = -0.6771; \quad \cos \theta = 0.7359$$

The procedure to obtain the rotational matrix is to place:

$$-\sin \theta \text{ at } a_{ij}; \quad \sin \theta \text{ at } a_{ji}; \quad \cos \theta \text{ at } a_{ii} \text{ and } a_{jj}$$

Hence:

$$R_1 = \begin{bmatrix} \cos \theta & -\sin \theta & 0 \\ \sin \theta & \cos \theta & 0 \\ 0 & 0 & 1 \end{bmatrix} = \begin{bmatrix} 0.7359 & 0.6771 & 0 \\ -0.6771 & 0.7359 & 0 \\ 0 & 0 & 1 \end{bmatrix}$$

a_{12} is now reduced to zero by the multiplication:

$$R_1{}^T[A]R_1 = \begin{bmatrix} 0.7359 & -0.6771 & 0 \\ 0.6771 & 0.7359 & 0 \\ 0 & 0 & 1 \end{bmatrix}\begin{bmatrix} 3 & 6 & 1 \\ 6 & 4 & 0 \\ 1 & 0 & 2 \end{bmatrix}\begin{bmatrix} 0.7359 & 0.6771 & 0 \\ -0.6771 & 0.7359 & 0 \\ 0 & 0 & 1 \end{bmatrix}$$

Hence:

$$B = \begin{bmatrix} -2.5208 & 0.0 & 0.7359 \\ 0.0 & 9.5208 & 0.6771 \\ 0.7359 & 0.6771 & 2.0 \end{bmatrix}$$

Second iteration
Largest valued off-diagonal element is now a_{13}.

$$\tan 2\theta = \frac{2 \times 0.7359}{-2.5208 - 2} = -0.3256; \quad \sin \theta = -0.1567; \quad \cos \theta = 0.9876$$

Hence:

$$R_2 = \begin{bmatrix} 0.9876 & 0 & 0.1567 \\ 0.00 & 1 & 0.00 \\ -0.1567 & 0 & 0.9876 \end{bmatrix}$$

and B now becomes:

$$\begin{bmatrix} 0.9876 & 0 & -0.1567 \\ 0.0 & 1 & 0.00 \\ 0.1567 & 0 & 0.9876 \end{bmatrix}\begin{bmatrix} -2.5208 & 0.0 & 0.7359 \\ 0.0 & 9.5208 & 0.6771 \\ 0.7359 & 0.6771 & 2.0 \end{bmatrix}\begin{bmatrix} 0.9876 & 0 & 0.1567 \\ 0.0 & 1 & 0.0 \\ -0.1567 & 0 & 0.9876 \end{bmatrix}$$

$$= \begin{bmatrix} -2.6376 & -0.1061 & 0.0 \\ -0.1061 & 9.5208 & 0.6687 \\ 0.0 & 0.6687 & 2.1168 \end{bmatrix}$$

and we see that a_{12} has changed from 0.0 to -0.1061, a value which will eventually have to be brought back to zero.

In all it takes five iterations to create off-diagonal values that have absolute values equal to or less than 0.0001. Even for this small order matrix the iterations involve so much computation that it is virtually impossible to carry them out manually and the use of a computer is essential. A summary of the whole operation is set out in Table 7.1

Table 7.1

n	Matrix			Rotation Matrix		
1	3.0000	6.0000	1.0000	0.7359	0.6771	0.0000
	6.0000	4.0000	0.0000	−0.6771	0.7359	0.0000
	1.0000	0.0000	2.0000	0.0000	0.0000	1.0000
2	−2.5208	0.0000	0.7359	0.9876	0.0000	0.1567
	0.0000	9.5208	0.6771	0.0000	1.0000	0.0000
	0.7359	0.6771	2.0000	−0.1567	0.0000	0.9876
3	−2.6376	−0.1061	0.0000	1.0000	0.0000	0.0000
	−0.1061	9.5208	0.6687	0.0000	0.9960	−0.0892
	0.0000	0.6678	2.1168	0.0000	0.0892	0.9960
4	−2.6376	−0.1057	0.0095	1.0000	−0.0086	0.0000
	−0.1057	9.5807	0.0000	0.0086	1.0000	0.0000
	0.0095	0.0000	2.0569	0.0000	0.0000	1.0000
5	−2.6385	0.0000	0.0095	1.0000	0.0000	−0.0020
	0.0000	9.5816	−0.0001	0.0000	1.0000	0.0000
	0.0095	−0.0001	2.0569	0.0020	0.0000	1.0000
6	−2.6385	0.0000	0.0000			
	0.0000	9.5816	−0.0001			
	0.0000	−0.0001	2.5069			

The eigenvalues are −2.6385, 9.5816 and 2.5069.

Eigenvectors by Jacobi's method

Without going into the theory it can be said that the normalised modal matrix, N, of matrix A is simply the product of the rotational matrices used to determine the eigenvalues of A, i.e.

$$N = R_1 . R_2 . R_3 \ldots R_n$$

It must be noted with Jacobi's method that the first column of the modal matrix, i.e. the eigenvector, corresponds to the eigenvalue in the first column of the spectral matrix, the two second columns similarly correspond and so on.

Example 7.12

Determine the eigenvectors corresponding to the eigenvalues 31.25 and 18.75 obtained in example 7.10.

Solution

In example 7.10 the rotational angle, θ, was found to be 45°.

Hence, $\sin \theta = 0.7071$ and $\cos \theta = 0.7071$ and the rotational matrix to eliminate a_{12} was therefore:

$$R = \begin{bmatrix} 0.7071 & -0.7071 \\ 0.7071 & 0.7071 \end{bmatrix}$$

As there was only one transformation R must also be equal to the normalised modal matrix N and the first column of this matrix is the eigenvector corresponding to b_{11}, i.e. 31.25.

Hence, for the eigenvalue 31.25, the eigenvector is

$$\begin{bmatrix} 0.7071 \\ 0.7072 \end{bmatrix}$$

and for the eigenvalue 18.75, the eigenvector is

$$\begin{bmatrix} -0.7071 \\ 0.7071 \end{bmatrix}$$

As a matter of interest the normalised modal matrix for example 7.11 is:

$$N = \begin{bmatrix} 0.7326 & 0.6784 & 0.0559 \\ -0.6621 & 0.7292 & -0.1727 \\ -0.1579 & 0.0895 & 0.9834 \end{bmatrix}$$

Further reading matter

In order to keep this chapter to a sensible size the material has been severely restricted to that necessary for the application of matrix algebra to statistics.

For the reader who would like to know more of the subject there is an excellent book written by Jennings (1977) which deals comprehensively with matrix algebra and its application to engineering problems.

Chapter Eight

Correlated and Non-normal variables

Multivariate distributions

Up to this point we have only considered the distributions of the values of single variables, i.e. variables unrelated to other variables. A set of values obtained for a single variable is usually referred to as univariate data.

When the values of one variable are affected by the values of another variable then a joint probability distribution relating these two variables must exist. Such a distribution is known as a bivariate distribution.

If there are several related random variables, the joint probability distribution is called a multivariate distribution.

The scatter diagram

An introduction to the treatment of multivariate data is probably best obtained from the study of bivariate data which, as it involves a two-dimensional probability function, can be represented by a diagram known as a scatter diagram.

An example of bivariate data where, for each value of the variable X there is a corresponding value of the variable Y, is set out below.

Table 8.1

x	0	1	2	3	4	5	6	7	8
y	2	1	5	3	1	6	9	5	4

With any bivariate data two unique straight lines exist:

(i) The regression line of Y on X.

Considering Y as the dependent variable the regression line of Y on X is the line that best estimates the value of Y corresponding to a value of X.

(ii) The regression line of X on Y.

Considering X as the dependent variable the regression line of X on Y is the line that best estimates the value of X corresponding to a value of y.

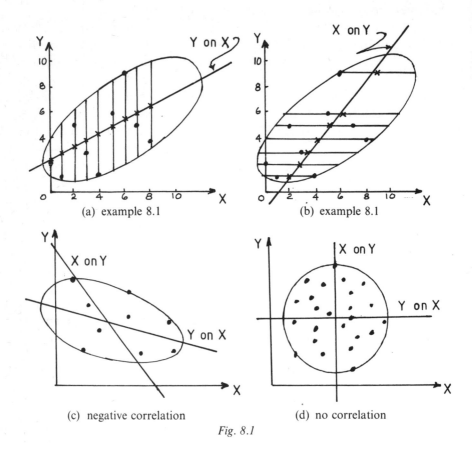

(a) example 8.1

(b) example 8.1

(c) negative correlation

(d) no correlation

Fig. 8.1

The scatter diagram shown in Figs 8.1a and b is a graphical illustration of the information contained in Table 8.1 and is formed by plotting the values as paired coordinates, (x, y), viz. $(0, 2), (1, 1), (2, 5), \ldots, (8, 4)$, OX being the horizontal axis. It is seen that the plotted points can all be enclosed within an ellipse.

To obtain the regression line of Y on X a vertical line is first drawn through each of the plotted points. The portion of each line enclosed within the ellipse is a vertical chord of the ellipse and its midpoint is marked. The best straight line through these points is the regression line of Y on X (see Fig. 8.1a).

Similarly the straight line that best joins the midpoints of the horizontal chords, obtained by drawing the set of horizontal lines through the plotted points, is the regression line of X on Y (see Fig. 8.1b).

Both of these lines slope upwards from left to right, i.e. the value of each variable increases as the value of the other increases and the variables are said to be positively correlated.

When a regression line slopes downwards from left to right, i.e. when the value of the dependent variable decreases as the value of the other variable increases, the variables are said to be negatively correlated. An illustration of negative correlation is shown in Fig. 8.1c.

When the plotted points of a scatter diagram can only be enclosed within a circle the two regression lines, X on Y and Y on X, are at right angles to each other and there is no correlation between X and Y (see Fig. 8.1d).

By studying Fig. 8.1 it can be deduced that there is complete correlation between the two variables when the two regression lines coincide whereas there is no correlation when the regression lines are at right angles to each other.

Mathematical determination of regression lines

There is always a risk of error when estimating by eye and, for most situations, it is best to use a mathematical approach which removes the risk of bias. The method of least squares is possibly the most popular and will be described here.

The regression line of Y on X is a straight line with an equation of the form:

$$y_{ei} = a_1 x_i + b_1$$

where y_{ei} is on the regression line and is the estimated value of Y when $X = x_i$.

The regression line of X on Y has a similar equation:

$$x_{ei} = a_2 y_i + b_2$$

The procedures for determining these two equations are identical and only the regression line of Y on X will therefore be discussed.

Unless there is perfect correlation between the two variables the regression line will not go through all the plotted points and there will be some errors of prediction.

For X equal to x_i the error in the prediction of the corresponding value of Y will be $y_i - y_{ei}$ where y_i is the actual observed value of Y.

It has been found that the regression line fits closest to the observed values when $S_{X,Y}$, the sum of the squares of the errors of the estimates, is a minimum. For n observed values:

$$S_{X,Y} = \sum_{i=1}^{n} (y_i - y_{ei})^2$$

Now, from the regression line equation:

$$y_i - y_{ei} = y_i - a_1 x_i - b_1$$

Therefore:

$$S_{X,Y} = \sum_{i=1}^{n} (y_i - a_1 x_i - b_1)^2$$

which is a function of a_1 and b_1: $f(a_1, b_1)$.

For this expression to be a minimum:

$$\frac{\partial f(a_1, b_1)}{\partial a_1} = \frac{\partial f(a_1, b_1)}{\partial b_1} = 0$$

$$\frac{\partial f(a_1, b_1)}{\partial a_1} = \sum_{i=1}^{n} \frac{\partial}{\partial a_1} (y_i - a_1 x_i - b_1)^2$$

$$= \sum_{i=1}^{n} -2x_i(y_i - a_1 x_i - b_1)$$

$$\frac{\partial f(a_1, b_1)}{\partial b_1} = \sum_{i=1}^{n} \frac{\partial}{\partial b_1} (y_i - a_1 x_i - b_1)^2$$

$$= \sum_{i=1}^{n} -2(y_i - a_1 x_i - b_1)$$

And, as both the above equations are equal to zero we finish with:

$$\sum_{i=1}^{n} (a_1 x_i + b_1 - y_i) = 0$$

Therefore:

$$\sum y_i = a_1 \sum x_i + nb_1 \qquad (1)$$

Therefore:

$$\sum_{i=1}^{n} (a_1 x_i + b_1 - y_i) = 0$$

Therefore:

$$\sum x_i y_i = a_1 \sum x_i^2 + b_1 \sum x_i \qquad (2)$$

Solving equations (1) and (2) gives the values of a_1 and b_1 which correspond to the minimum value of $S_{X',Y}$.

Example 8.1

Using the tabulated data in Table 8.1 determine the regression line of Y on X.

Solution
Examining equations (1) and (2) derived in the preceding section it is seen that they include the terms x_i, y_i, $x_i y_i$, and x_i^2.

There is less likelihood of error if the calculations to determine the values of these terms are tabulated:

x_i	y_i	x_i^2	$x_i y_i$
0	2	0	0
1	1	1	1
2	5	4	10
3	3	9	9
4	1	16	4
5	6	25	30
6	9	36	54
7	5	49	35
8	4	64	32
$\sum 36$	$\sum 36$	$\sum 204$	$\sum 175$

Equation (1) is therefore:

$$36 = 36a_1 + 9b_1$$

and equation (2) is:

$$175 = 204a_1 + 36b_1$$

Solution of these equations gives a_1 as 0.517 and b_1 as 1.932. The regression lines equation for Y on X is:

$$y_{ei} = 0.517x_i + 1.932$$

The reader might like to check that the equation of the regression line of X on Y is:

$$x_{ei} = 0.574y_i + 1.704$$

The two lines are shown in Figs 8.1a and b.

Joint probability functions

As mentioned above the joint behaviour of two or more correlated random variables, is described by their joint probability function, which, depending on whether the variables are discrete or continuous, will be either a joint probability mass function or a joint probability density function.

JOINT pmf

The joint probability mass function of two discrete random variables, X and Y, is the expression for $p_{X,Y}(x, y)$, the probability that X will achieve a value x and that Y will achieve a value y.

$$p_{X,Y}(x, y) = P[(X = x) \cap (Y = y)]$$

JOINT pdf

The probability functions of continuous variables are virtually the same as for discrete variables except that the summation signs are replaced by integrals the mass functions being replaced by density functions.

The joint probability density function of two continuous random variables, X and Y, is written as $f_{X,Y}(x, y)$ and is a shorthand expression for the probability that X lies within the range from x to $x + dx$ whilst Y lies within the range from y to $y + dy$.

The function $f_{X,Y}(x, y)$ obeys the probability laws:

$$f_{X,Y}(x, y) = 0$$

$$\int_{-\infty}^{\infty} \int_{-\infty}^{\infty} f_{X,Y}(x, y)\, dx\, dy = 1$$

The function $Z = f_{X,Y}(x, y)$ is the equation of a surface, usually called the probability surface.

The volume of the space contained within the probability surface and the X, Y plane is equal to one (as can be seen from the second law above).

The probability that X lies within the range x_1 to x_2 whilst Y lies within a range y_1 to y_2 can be expressed mathematically as:

$$P[(x_1 \leqslant X \leqslant x_2) \cap (y_1 \leqslant Y \leqslant y_2)]$$

which is equal to:

$$\int_{x_1}^{x_2} \int_{y_1}^{y_2} f_{X,Y}(x, y)\, dx\, dy$$

Obviously the more complicated the joint probability density function, i.e. the equation of the probability surface, the more complicated is the integration necessary to evaluate probability values and the problem becomes one of mathematics rather than of probability theory. Fortunately the joint behaviour of related random variables can usually be predicted from their moments in an exactly similar manner to the use of the mean and variance of a single variable when its exact probability distribution is either not known or is difficult to integrate.

A knowledge of this aspect of jointly distributed variables is all that is required in order to make use of the reliability analysis method described in this book and, because of this, the study of joint probability functions will not be taken further.

Covariance

In chapter 2 the variance of single random variables was discussed and it was shown that, for a variable X, its variance could be obtained from the expression:

$$\text{Var}(X) = E[(X - m_X)^2]$$
$$= \sum_{i=1}^{n} \frac{(x_i - m_X)^2}{n}$$

for a statistical sample of n values.

In a similar manner, if Z is a function of two related random variables X and Y so that Z equals $g(X, Y)$ then the covariance of Z, $\text{COV}(Z)$ is found from the expression:

$$\text{COV}(Z) = \text{COV}(X, Y) = E[X - m_X][Y - m_Y]$$
$$= \sum_{i=1}^{n} \frac{(x_i - m_X)(y_i - m_Y)}{n}$$

for a statistical sample of n related values.

The computational effort is reduced if this latter formula is rewritten as:

$$\text{COV}(X, Y) = \frac{1}{n}\left[\sum x_i y_i - \frac{\sum x_i \sum y_i}{n}\right]$$

Liner correlation coefficient

This important term, given the symbol r, is a means of expressing numerically just how closely the regression lines of two related variables X and Y fit the observed values.

The degree of correlation, r, has the same value for both regression lines and is a normalised version of the covariance. It can be obtained from the formula:

$$r = \frac{COV(X, Y)}{\sigma_X \sigma_Y}$$

If r is negative there is a negative correlation and Y decreases in value as X increases whereas if r is positive Y increases as X increases and there is positive correlation.

r can be either positive or negative with a value ranging from -1 to $+1$. If r equals zero then there is no correlation between X and Y and if r equals one there is perfect correlation between the two variables.

Full correlation means that, for any given value of X, the value of Y estimated from the regression line, $y(e)$, will be equal to the observed value, y.

The following is a rough guide as to the interpretation of the numerical value of r:

$\|r\| = 0.8$	Strong correlation between X and Y which can be assumed to be completely dependent.
$0.8 > \|r\| > 0.2$	Correlation between X and Y
$\|r\| = 0.2$	Weak correlation between X and Y which can be assumed independent of each other.

As the level II method of reliability analysis approximates the failure boundary to a linear surface, we need only deal with linear correlation in these notes. However it should be remembered that, although an $|r|$ value less than 0.2 indicates little linear correlation between X and Y it may well be that there is a strong correlation of some non-linear form.

Standard error of the estimate

The variance of a random variable is a measure of the amount of variability about the mean value of the variable.

In a similar manner the sum of the squares of the estimation errors, S, is a measure of the variability of the observed values about the regression line. If the regression line is the Y on X line then the sum should be given the symbols $S_{Y,X}$ and it is simply the variance of Y on X.

The standard deviation of a variable is the square root of its variance and, in a similar manner, the standard error of the estimate is the square root of the variance of Y on X and is given the symbol $s_{Y,X}$.

There is also a standard error of the estimate $s_{X,Y}$ because there are two regression lines, one for estimating Y values and one for estimating X values.

A regression line gives the best estimated value possible from the data available but it is still only an estimate and the value of the standard error of this estimate is a useful guide as to the accuracy of the predicted value.

Assuming a normal distribution we can say that just over 95% of the actual values will be within plus or minus two standard errors of the estimated values. We can draw these lines on the regression line diagram to give a visual representation of the confidence one should have in the estimated values. The 95% control lines for example 8.1 are shown in Fig. 8.2.

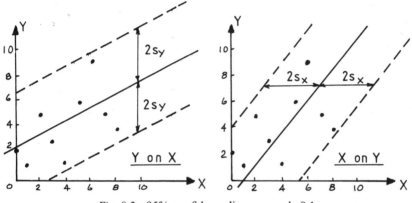

Fig. 8.2   95% confidence lines example 8.1

It can be shown that the standard errors are related to the linear correlation coefficient by the equations:

$$s_{Y,X} = \sigma_Y\sqrt{(1 - r^2)}$$
$$s_{X,Y} = \sigma_X\sqrt{(1 - r^2)}$$

Example 8.2

Using the data of example 8.1 determine: (i) the covariance of X and Y; (ii) the correlation coefficient; and (iii) the standard errors of the estimates.

Solution
(i)

$$\text{COV}(X, Y) = \sum \frac{(x_i - m_X)(y_i - m_Y)}{n}$$

$$= \tfrac{1}{9}\left(175 - \frac{36 \times 36}{9}\right) = 3.444$$

(ii)

$$r = \frac{COV(X, Y)}{\sigma_X \cdot \sigma_Y} = \frac{3.444}{2.582 \times 2.450} = 0.544$$

(iii)

$$s_{Y,X} = \sigma_Y \sqrt{(1 - r^2)} = 2.450(1 - 0.544^2) = 2.055$$

$$s_{X,Y} = \sigma_X \sqrt{(1 - r^2)} = 2.582(1 - 0.544^2) = 2.165$$

In this example the value of r shows that there is some correlation between X and Y but it is fairly weak, as indicated in Figs 8.1a and b.

The Bessel correction

As discussed in chapter 2 the amount of data available for most civil engineering statistical samples is small and the standard deviation of the variable is best estimated using a denominator of $(n - 1)$ instead of n. It is also realistic to apply this correction to the covariance values of civil engineering variables.

If Bessel's correction had been applied in example 8.2 then the COV(X, Y) would have been equal to $3.444 \times \frac{9}{8} = 3.875$ although the value of r would remain at 0.544 as σ_X and σ_Y would have increased to 2.739 and 2.598 respectively.

Treatment of correlated variables

The level II method of reliability analysis is based on the assumption that the variables in the limit state equation are statistically independent, i.e. they are not correlated.

For weakly correlated variables, $|r|$ equal to 0.2, the assumption of no correlation is usually satisfactory whereas for $|r|$ values equal to 0.8 complete dependence can be assumed, effectively reducing the number of variables by one.

For other related variables the effect of correlation can be allowed for and, when possible, it is probably best to do so although work carried out by the author indicates that, when several variables are involved there is often a cancelling-out effect and the value obtained for β is not very different to the value obtained when full independence of the variables is assumed.

The covariance matrix

The procedure to allow for correlation effects uses the covariance matrix to transform the space of the correlated variables into a space where there is no correlation between them.

The covariance matrix is the matrix which lists the correlations between the correlated variables and is illustrated in Table 8.2 for a set of related variables $X_1, X_2, X_3, \ldots, X_n$.

Table 8.2

$$
\begin{array}{c}
\begin{array}{cccc} X_1 & X_2 & X_3 & X_4 \ldots X_n \end{array} \\
\begin{array}{c} X_1 \\ X_2 \\ X_3 \\ X_4 \\ \vdots \\ X_{n-1} \\ X_n \end{array}
\left[
\begin{array}{ccccc}
\sigma_{X_1}{}^2 & \mathrm{COV}(X_1, X_2) & \mathrm{COV}(X_1, X_3) & \mathrm{COV}(X_1, X_4) \ldots \mathrm{COV}(X_1, X_n) \\
\mathrm{COV}(X_2, X_1) & \sigma_{X_2}{}^2 & \mathrm{COV}(X_2, X_3) & \mathrm{COV}(X_2, X_4) \ldots \mathrm{COV}(X_2, X_n) \\
\mathrm{COV}(X_3, X_1) & \mathrm{COV}(X_3, X_2) & \sigma_{X_3}{}^2 & \mathrm{COV}(X_3, X_4) \ldots \mathrm{COV}(X_3, X_n) \\
\vdots & \vdots & \vdots & \vdots \qquad\qquad \vdots \\
\mathrm{COV}(X_{n-1}, X_1) & \ldots & \mathrm{COV}(X_{n-1}, X_{n-2}) \; \sigma_{X_{n-1}}{}^2 & \mathrm{COV}(X_{n-1}, X_n) \\
\mathrm{COV}(X_n, X_1) & \mathrm{COV}(X_n, X_2) & \mathrm{COV}(X_n, X_3) & \cdots \qquad \sigma_{X_n}{}^2
\end{array}
\right]
\end{array}
$$

The leading diagonal is made up of the values of the variances of the correlated variables and the off-diagonal terms are the corresponding covariances.

The space transformation is achieved by an orthogonal transformation devised by Hasofer and Lind (1974), in which the axes defining the original variables are lined up parallel with the eigenvectors of the covariance matrix.

Essentially the method consists of determining the eigenvalues and eigenvectors of the covariance matrix. This can be achieved by diagonalising the matrix using Jacobi's method, which is described in chapter 7. Once the transformation has been achieved the eigenvalues are the elements on the leading diagonal and, knowing these values, the corresponding eigenvectors can be obtained.

The purpose of the transformation is that the eigenvalues are the variances, and the eigenvectors when multiplied by the original variables, $X_1, X_2, \ldots, X_n$, are the values, of a set of equivalent but transformed and uncorrelated variables $Y_1, Y_2, \ldots, Y_n$. These uncorrelated variables are now normalised to give a set of reduced uncorrelated variables $y_1, y_2, \ldots, y_n$ corresponding to the original variables $X_1, X_2, \ldots, X_n$ from which the reliability index can be found.

The procedure is illustrated in the following examples.

Example 8.3

A simply supported beam has a span, L, of 9 m and supports two vertical loads, P_1 and P_2, at its third points as shown in Fig. 8.3.

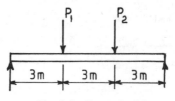

Fig. 8.3   Example 8.3

Young's modulus for the beam, E_b, is 20 000 MN/m² and its moment of inertia, I_b, is equal to 0.015 m⁴.

Both P_1 and P_2 have mean values of 30 kN and standard deviations of 5 kN whilst the values of L, E_b, and I_b are constant.

If the allowable deflection at the centre of the span is 30 mm, determine the probability that this value will be exceeded:

(i) If P_1 and P_2 are uncorrelated.
(ii) If P_1 and P_2 have a coefficient of correlation equal to 0.25.

Solution
(i) All variables uncorrelated.

R = allowable deflection = 30 mm

$$S = \text{deflection caused by loads} = \frac{kL^3P_1}{E_bI_b} + \frac{kL^3P_2}{E_bI_b}$$

Now $k = 0.1728$ (Reynolds, 1957) hence, $S = 0.4199P_1 + 0.4199P_2$
and:

$$Z = 30 - 0.4199P_1 - 0.4199P_2$$

$$= 30 - 0.4199(y_1\sigma_1 + m_1) - 0.4199(y_2\sigma_2 + m_2)$$

The iterative procedure for β gives:

	y_1	y_2	β	$h(y)$
1	0.0	0.0	0.0	4.81
2	1.15	1.15	1.62	0.00

With the assumption of no correlation, $\beta = 1.62$

(ii) $r = 0.25$

$$(\sigma_{X_1})^2 = (\sigma_{X_2})^2 = 25$$

$$\text{COV}(X, Y) = \text{COV}(Y, X) = r\sigma_{X_1}\sigma_{X_2} = 6.25$$

Hence the covariance matrix is:

$$\begin{bmatrix} 25 & 6.25 \\ 6.25 & 25 \end{bmatrix}$$

Examples 7.10 and 7.12 have illustrated that this matrix, when transformed into a diagonal matrix is:

$$\begin{bmatrix} 31.25 & 0.0 \\ 0.0 & 18.75 \end{bmatrix}$$

with the transposed modal matrix given by

$$N^T = \begin{bmatrix} 0.7071 & 0.7071 \\ -0.7071 & 0.7071 \end{bmatrix}$$

Hence the transformed variables, Y_1, equivalent to X_1, and Y_2, equivalent to X_2, are:

$$Y_1 = 0.7071X_1 + 0.7071X_2 \quad \text{with } \sigma_{Y_1} = \sqrt{31.25} = 5.59$$

$$Y_2 = -0.7071X_1 + 0.7071X_2 \quad \text{with } \sigma_{Y_2} = \sqrt{18.75} = 4.33$$

Now:

$$E[Y_1] = 0.7071E[X_1] + 0.7071E[X_2] = 0.7071(30 + 30) = 42.43$$

and:

$$E[Y_2] = -0.7071E[X_1] + 0.7071E[X_2] = 0$$

Now a reduced variable, y, is equal to $(y - m_Y)/\sigma_Y$. Hence:

$$y_1 = \frac{0.7071X_1 + 0.7071X_2 - 42.43}{5.59}$$

$$= 0.1265X_1 + 0.1265X_2 - 7.59 \tag{1}$$

and:

$$y_2 = \frac{-0.7071X_1 + 0.7071X_2 - 0}{4.33}$$

$$= -0.1633X_1 + 0.1633X_2 \tag{2}$$

Solving equations (1) and (2) for X_1 and X_2 gives:

$$X_1 = 3.954y_1 + 3.063y_2 + 30.01$$

$$X_2 = 3.954y_1 - 3.063y_2 + 30.01$$

Now:

$$Z = 30 - 0.4199X_1 - 0.4199X_2$$
$$= 30 - 0.4199(7.908y_1 + 60.02)$$
$$= 4.798 - 3.321y_1$$

The iterative procedure gives β as $4.798/3.321 = 1.45$.

The relationship between β and r for values of r ranging from -0.75 to 0.75 is shown plotted in Fig. 8.4 and illustrates that, for this example at least, the value of r has a pronounced effect on the value of β as only two related variables are involved.

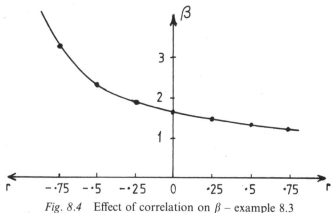

Fig. 8.4 Effect of correlation on β – example 8.3

Example 8.4

A continuous foundation, 2 m wide, is to be placed on the surface of a granular soil and will support a uniform vertical pressure, p. The uniform pressure induced by the weight of the foundation will be 10 kN/m^2 which, along with the 2 m dimension, may be assumed to be constant.

The applied pressure, p, will have a mean value of 400 kN/m^2 and a standard deviation of 30 kN/m^2.

A series of shear tests carried out on a set of random samples taken from the soil gave the values of unit weight, γ, and the corresponding angles of shearing resistance, ϕ, listed below.

γ (kN/m³)	18.0	18.8	19.2	19.4	19.5	19.7	20.1	20.2
ϕ (degrees)	33.3	33.5	34.8	33.1	36.8	35.4	35.5	37.0

Determine the probability of failure of the foundation:

(i) Assuming that the unit weight and the angle of shearing resistance of the soil are uncorrelated.

(ii) Allowing for any correlation between the unit weight and the angle of shearing resistance.

Solution
Using the formula for N_γ, given in chapter 5, the following table can be completed:

γ (kN/m³)	18.0	18.8	19.2	19.4	19.5	19.7	20.1	20.2
N_γ	25.67	26.52	32.82	24.84	45.81	36.24	36.85	47.38

Using the processes described in the beginning of this chapter, or by means of a suitable pocket calculator, the values listed below can be established. It should be noted that, due to the small number of values in the sample, the Bessel correction has been applied.

$$m_\gamma = 19.36; \qquad (\sigma_\gamma)^2 = 0.511; \qquad \sigma_\gamma = 0.72$$
$$m_{N_\gamma} = 34.52; \qquad (\sigma_{N_\gamma})^2 = 77.164; \qquad \sigma_{N_\gamma} = 8.78$$
$$COV(\gamma, N_\gamma) = 4.442; \qquad r = 0.707$$

(i) Assuming no correlation

$$Q_u = 0.5B\gamma N_\gamma = 0.5 \times 2.0\gamma N_\gamma = \gamma N_\gamma = R$$
$$S = p + \text{pressure due to foundation} = p + 10$$

Hence:
$$Z = \gamma N_\gamma - p - 10$$

Putting $\gamma = X_1$, $N_\gamma = X_2$ and $p = X_3$ gives:
$$Z = X_1.X_2 - X_3 - 10$$

and, expressing Z in terms of reduced variables:

$$Z = (y_1\sigma_1 + m_1)(y_2\sigma_2 + m_2) - (y_3\sigma_3 + m_3)$$
$$= (0.72y_1 + 19.36)(8.78y_2 + 34.52) - (30y_3 + 400) - 10$$
$$= (0.72y_1 + 19.36)(8.78y_2 + 34.52) - 30y_3 - 410$$

Differentiating gives:
$$g_1' = 0.72(8.78y_2 + 34.52)$$
$$g_2' = 8.78(0.72y_1 + 19.36)$$
$$g_3' = -30$$

The iterative procedure to obtain β gives the folowing results:

	y_1	y_2	y_3	β	$h(y)$
1	0.0	0.0	0.0	0.0	258.31
2	-0.21	-1.44	0.26	1.48	1.93
3	-0.14	-1.46	0.26	1.49	-0.35
4	-0.14	-1.46	0.26	1.49	0.00

(ii) Allowing for correlation
The covariance matrix is:

$$\begin{bmatrix} 0.511 & 4.442 \\ 4.442 & 77.164 \end{bmatrix}$$

This matrix can be transformed into a diagonal matrix, as discussed in chapter 7, and becomes:

$$\begin{bmatrix} 0.2544 & 0 \\ 0 & 77.42 \end{bmatrix}$$

with the corresponding normalised modal matrix, N:

$$\begin{bmatrix} 0.9983 & 0.0577 \\ -0.0577 & 0.9983 \end{bmatrix}$$

The uncorrelated variables Y_1 and Y_2 corresponding to X_1 and X_2 can now be found from the relationship $Y = N^T X$:

$$Y_1 = 0.9983X_1 - 0.0577X_2 \quad \text{with } \sigma_{Y_1} = \sqrt{0.2544} = 0.504$$

$$Y_2 = 0.0577X_1 + 0.9983X_2 \quad \text{with } \sigma_{Y_2} = \sqrt{77.42} = 8.80$$

Now:

$$E[Y_1] = N^T E[X] = 0.9983E[X_1] - 0.0577E[X_2]$$

$$= 0.9983 \times 19.36 - 0.0577 \times 34.52 = 17.34$$

$$E[Y_2] = 0.0577 \times 19.36 + 0.9983 \times 34.52 = 35.58$$

Reduced, or standardised, variables have been discussed in chapter 4 and are obtained by an equation of the form:

$$y = \frac{Y - E[Y]}{\sigma_Y}$$

Hence:

$$y_1 = \frac{0.9983X_1 - 0.0577X_2 - 17.34}{0.504}$$

$$= 1.8080X_1 - 0.1145X_2 - 34.405$$

and:

$$y_2 = \frac{0.0577X_1 + 0.9983X_2 - 35.58}{8.80}$$

$$= 0.00656X_1 + 0.1134X_2 - 4.043$$

Solving these two equations gives:

$$X_1 = 0.503y_1 + 0.508y_2 + 19.37$$
$$X_2 = -0.032y_1 + 8.786y_2 + 34.61$$

Now:

$$Z = X_1X_2 - X_3 - 10$$

and, substituting for X_1, X_2 and X_3 gives:

$$Z = (0.503y_1 + 0.508y_2 + 19.37)(-0.032y_1 + 8.786y_2 + 34.61)$$
$$- 30y_3 - 400 - 10$$
$$= -0.16y_1{}^2 + 16.79y_1 + 4.4y_1y_2 + 187.76y_2$$
$$+ 4.46y_2{}^2 - 30y_3 + 260.4$$

Differentiating Z gives:

$$g_1' = -0.32y_1 + 16.79 + 4.4y_2$$
$$g_2' = 4.4y_1 + 375.52y_2 + 4.46$$
$$g_3' = -30$$

The iterative procedure for β is set out below:

	y_1	y_2	y_3	β	$h(y)$
1	0.0	0.0	0.0	0.0	260.41
2	-0.12	-1.34	0.21	1.36	8.74
3	-0.09	-1.39	0.24	1.41	-0.15
4	-0.09	-1.39	0.24	1.41	0.00

Hence, when correlation effects are allowed for, β is 1.41 but, if no correlation is assumed, β is 1.49.

Example 8.5

The limit state equation for a particular failure mechanism is:

$$Z = 50 + 5X_1 - 4X_2 - 2X_3 + X_4$$

The variables have the following values:

Variable	Mean	s.d.
X_1	15	3.5
X_2	10	3.0
X_3	15	2.5
X_4	5	1.5

The variables X_1, X_2 and X_3 are correlated and the covariance matrix is:

$$
\begin{array}{c}
\quad X_1 \quad X_2 \quad X_3 \\
\begin{array}{c} X_1 \\ X_2 \\ X_3 \end{array}
\begin{bmatrix}
12.25 & -5 & 2 \\
-5 & 9 & -4 \\
2 & -4 & 6.25
\end{bmatrix}
\end{array}
$$

Determine the reliability index of the system: (i) assuming no correlation effects; and (ii) allowing for correlation.

Solution
(i) Assuming no correlation

$$Z = 50 + 5X_1 - 4X_2 - 2X_3 + X_4$$

Expressing in reduced variables:

$$Z = 50 + 5(3.5y_1 + 15) - 4(3y_2 + 10) - 2(2.5y_3 + 15) + (1.5y_4 + 5)$$

The differentiations are straightforward and the iterative procedure for β gives:

	y_1	y_2	y_3	y_4	β	$h(y)$
1	0.0	0.0	0.0	0.0	0.0	60.0
2	-2.20	1.51	0.63	-0.19	2.75	0.00
3	-2.20	1.51	0.63	-0.19	2.75	0.00

(ii) Allowing for correlation
The covariance matrix can be transformed into a diagonal matrix using a technique such as the Jacobi method which is described in chapter 7. The

transformed matrix is:

$$\begin{bmatrix} 17.220 & 0.0 & 0.0 \\ 0.0 & 6.198 & 0.0 \\ 0.0 & 0.0 & 3.112 \end{bmatrix}$$

Hence:

$$\begin{bmatrix} \sigma_{Y_1} = 4.150 \\ \sigma_{Y_2} = 2.490 \\ \sigma_{Y_3} = 1.764 \end{bmatrix}$$

and the corresponding normalised modal matrix, N, is:

$$N = \begin{bmatrix} 0.7144 & 0.6660 & 0.2147 \\ -0.6054 & 0.4343 & 0.6670 \\ 0.3510 & -0.6065 & 0.7134 \end{bmatrix}$$

Therefore:

$$Y_1 = 0.7144X_1 - 0.6054X_2 + 0.3510X_3 \quad \text{and } E[Y_1] = 9.927$$
$$Y_2 = 0.6660X_1 + 0.4343X_2 - 0.6065X_3 \quad \text{and } E[Y_2] = 5.235$$
$$Y_3 = 0.2147X_1 + 0.6670X_2 + 0.7134X_3 \quad \text{and } E[Y_3] = 20.592$$

Therefore:

$$y_1 = \frac{0.7144X_1 - 0.6054X_2 + 0.3510X_3 - 9.927}{4.150}$$

$$= 0.172X_1 - 0.146X_2 + 0.085X_3 - 2.392$$

and:

$$y_2 = 0.267X_1 + 0.174X_2 - 0.244X_3 - 2.102$$

and:

$$y_3 = 0.122X_1 + 0.378X_2 + 0.404X_3 - 11.673$$

Expressing in matrix form gives:

$$\begin{bmatrix} y_1 + 2.392 \\ y_2 + 2.102 \\ y_3 + 11.673 \end{bmatrix} = \begin{bmatrix} 0.172 & -0.146 & 0.085 \\ 0.267 & 0.174 & -0.244 \\ 0.122 & 0.378 & 0.404 \end{bmatrix} \begin{bmatrix} X_1 \\ X_2 \\ X_3 \end{bmatrix}$$

and, inverting the matrix, leads to:

$$\begin{bmatrix} X_1 \\ X_2 \\ X_3 \end{bmatrix} = \begin{bmatrix} 2.965 & 1.662 & 0.380 \\ -2.511 & 1.078 & 1.179 \\ 1.454 & -1.511 & 1.275 \end{bmatrix} \begin{bmatrix} y_1 + 2.392 \\ y_2 + 2.102 \\ y_3 + 11.673 \end{bmatrix}$$

Now:

$$Z = 50 + 5X_1 - 4X_2 - 2X_3 + X_4$$

and, substituting for X_1, X_2, X_3 and X_4, eventually gives:

$$Z = 60 + 21.96y_1 + 7.072y_2 - 5.33y_3 + 1.5y_4$$

and the iterative procedure for β gives:

	y_1	y_2	y_3	y_4	β	$h(y)$
1	0.0	0.0	0.0	0.0	0.0	60.00
2	-2.34	-0.75	0.57	-0.16	2.53	0.00
3	-2.34	-0.75	0.57	-0.16	2.53	0.00

It is seen that the value of the reliability index, when correlation effects are allowed for, is 2.53 whereas when correlation is ignored the value slightly increases to 2.75.

Correlation effects in soil mechanics

In structural engineering there will generally be enough dependable statistical information available to obtain reasonable estimates for the values of correlation coefficients and covariances. When this happens correlation can be allowed for as has been illustrated.

With soil mechanics there is never enough information and the designer is almost forced to assume that variables involved in a limit state function dealing with a soil failure mechanism are independent of each other. However, such an assumption is perhaps not all that radical.

Firstly when a number of variables is involved, the value of β obtained when correlation is allowed for is often fairly close to the value obtained when correlation is ignored.

Secondly the disturbance of the soil during construction indicates that the assumption of independence of the soil parameters is probably more realistic than any attempt to obtain some form of relationship between them.

Treatment of variables with non-normal distributions

If it is known that some of the variables involved in a reliability analysis have distributions that are not normal then the accuracy of the analysis will be increased if these variations are allowed for.

Rackwitz and Fiessler (1978), proposed a method for the treatment of independent non-normal random variables and it is the use of this method that will be described here. The authors showed that a non-normally distributed random variable, X_i, can be approximated at any point to a normally distributed one provided that both the cumulative density distributions (cdfs) and the probability density functions (pdfs) are equal at the point chosen. For the purpose of the second moment method of reliability analysis the point chosen is the design point, x_i^*, defined in chapter 4.

With these conditions the values of m_i^N and σ_i^N, the mean and standard deviation of the normalised variable, can be found from the expressions:

$$\sigma_i^N = f^N \frac{[\Phi^{-1}(F_{Xi}(x_i^*))]}{f_{Xi}(x_i)}$$

$$m_i^N = x_i^* - \Phi^{-1}(F_{Xi}(x_i^*))\sigma_i^N$$

where:

$F_{Xi}(x_i^*)$ = the cumulative probability of X_i at x_i^*
$f_{Xi}(x_i^*)$ = the probability density of X_i at x_i^*
$\Phi^{-1}(.)$ = the inverse normal distribution
$f^N(.)$ = the standardised normal density function

Values for the last two items are obtained from appendices II and III.

The values of m_i^N and σ_i^N are used instead of m_i and σ_i in the iteration procedure for determining β.

Example 8.6

For example 5.3, determine the probability that the allowable deflection will be exceeded if the point load, W, has an extreme type I distribution.

Solution

The mean and standard deviation of P are 800 and 125 kN respectively and the first step is to fit these values to an extreme distribution, as described in chapter 6. Now:

$$\alpha = \frac{1.282}{\sigma_Y} = \frac{1.282}{125} = 0.0103$$

and:

$$u = m_Y - \frac{0.577}{\alpha} = 800 - \frac{0.577}{0.0103} = 744$$

Using the expressions for $f_Y(y)$ and $F_Y(y)$ given in chapter 6:

$$f_Y(y) = 0.0033; \qquad F_Y(y) = 0.5703$$

Using Rackwitz and Fiessler's method we require first to determine the value of $\Phi^{-1}F_Y(y) = \Phi^{-1}(0.5703)$.

Now $\Phi^{-1}(y)$ is the inverse function so that, although we use appendix III, we go into the table and find the probability value 0.5703 and then find the z value corresponding to it. As the value 0.5703 is not actually tabulated some interpolation is necessary. It follows, by interpolation from appendix III:

$$\Phi^{-1}(0.5703) = 0.177$$

And, from appendix II:

$$f^N[\Phi^{-1}(0.5703) = f^N(0.177) = 0.3927$$

and thus:

$$\sigma_W{}^N = \frac{0.3927}{0.0033} = 119 \text{ kN}; \qquad m_W{}^N = 800 - 0.177 \times 119 = 779 \text{ kN}$$

hence the values of the means and standard deviations of the variables have become:

Variable	Symbol	Mean value	s.d.	Units
E	X_1	20 000	2 500	MN/m^2
I	X_2	0.01	0.0005	m^3
W	X_3	779	119	kN
k	X_4	0.1878	0.0239	

Now:

$$Z = 100(y_1\sigma_1 + m_1)(y_2\sigma_2 + m_2) - 64(y_3\sigma_3 + m_3)y_4\sigma_4 + m_4)$$

and the iterative procedure for β gives the following results:

	y_1	y_2	y_3	y_4	β	$h(y)$
1	0.0	0.0	0.0	0.0	0.0	10637.0
2	-2.48	-0.99	1.42	1.18	3.25	2.25
3	-2.33	-0.68	1.62	1.42	3.25	-114.1
4	-2.30	-0.67	1.61	1.41	3.21	-0.12
5	-2.30	-0.68	1.61	1.41	3.22	-0.01
6	-2.30	-0.68	1.61	1.41	3.22	0.00

The reliability index has increased from 3.09 to 3.22.

Example 8.7

Determine the probability of bearing capacity failure in example 5.11 if the bearing capacity coefficient N6 has a lognormal distribution.

Solution
The limit state equation was found to be:

$$Z = (y_1\sigma_1 + m_1)(y_2\sigma_2 + m_2) - (y_3\sigma_3 + m_3) - 12$$

The variables, and their input values, were:

Variable	Symbol	Mean	s.d.
γ	X_1	18	0.9
N_γ	X_2	79.54	14.05
p	X_3	500	30

If N_γ has a lognormal distribution then its mean and standard deviations must be transformed using the procedure described in chapter 6.

$$V = \frac{14.05}{79.54} = 0.1766; \qquad \sigma_X{}^2 = 0.0307 \quad \text{and} \quad \sigma_X = 0.1753$$

Hence:

$$\check{m}_Y = 78.33; \qquad z = 0.0875$$

From appendix II:

$$f_Z(z) = 0.3974$$

which gives:

$$f_Y(y) = 0.0285$$

Due to the fact that we establish the value of $F_Y(y)$ using the value of z, it is obvious that $\Phi^{-1}F_Y(y)$ must be equal to z and therefore $f^N(\Phi^{-1}F_Y(y))$ must be equal to $f_Z(z)$, i.e. 0.3974. Therefore:

$$\sigma_3{}^N = \frac{0.3974}{0.0285} = 13.94; \qquad m_3 = 79.54 - 0.0875 \times 13.94 = 78.31$$

The input parameters have therefore become:

Symbol	Mean	s.d.
X_1	18	0.9
X_2	78.31	13.94
X_3	500	30

and the iterative procedure for β gives:

y_1	y_2	y_3	β	$h(y)$
0.0	0.0	0.0	0.0	897.6
−0.92	−3.27	0.39	3.42	37.73
−0.43	−3.52	0.44	3.57	−10.95
−0.37	−3.49	0.43	3.53	0.08
−0.38	−3.49	0.42	3.53	−0.00

The reliability index is decreased slightly from 3.59 to 3.53.

Combination of different types of varying loads

The form of loading to which a structure can be subjected is extremely complex and, just as for a deterministic design, it must be modelled in some form before it can be included in a level II reliability analysis.

A parent distribution, such as $f_X(x)$, is often called the arbitrary point in time distribution of X or the first order distribution of X and example 6.4 illustrates how the parent distribution of a time-varying variable, X, can be used to determine the distribution of Y, the extreme value of X.

Being a function of time, an extreme load can always occur although the chance of it happening at all is extremely small. Obviously if there is a possibility during the lifetime of a structure that it could be subjected to an extreme load then it must be designed to withstand it. Such a situation presents no problem to the design engineer and the extreme value of the loading is included as part of the live loading.

However, if a structure is liable to be subjected to more than one single time-varying load during its operational life than it is very unlikely that these loads will achieve their extreme values at the same time and some economy in design can be obtained if this fact is taken into account. This can be achieved by assuming that each time varying load is independent of the others.

A simple load combination model which is still used was suggested by Turkstra (1970). The principle of the model is the assumption that when one time-varying load achieves its extreme value the other time-varying loads which are acting simultaneously, are at their instantaneous point in time values. For n time-varying loads there will be n limit state equations.

The model is best illustrated with an example.

Example 8.8

A simply supported beam has a span of 6 m and a moment of resistance M_f. The beam carries a uniform load, including its own weight, w_s, and a central point load is the sum of two loads, P_1 and P_2.

The details of the various design parameters are as follows:

Variable	Symbol	Mean	s.d.	Units
P_1	X_1	30	5	kN
P_2	X_2	40	5	kN
w_s	X_3	5	0.5	kN/m
M_f	X_4	1100	100	kNm

P_1 and P_2 are both time-varying loads and the values quoted are for weekly readings, taken over one year.

Determine the reliability index against bending failure for a time period of fifty years.

Solution

Using the Monte Carlo simulation method explained in chapter 6 it is possible to obtain the mean and standard deviation values for the weekly load values that can occur in a year. To obtain the mean and standard deviation of the extreme values for a period of 50 years the iteration procedure must be carried out 50 times. This task was carried out on a microcomputer with the following extreme values:

P_1 mean = 41.93 kN; s.d. = 2.73 kN

P_2 mean = 29.16 kN; s.d. = 2.19 kN

which transform to:

$$P_1^N \quad \text{mean} = 41.47 \text{ kN}; \quad \text{s.d.} = 2.61 \text{ kN}$$

$$P_2^N \quad \text{mean} = 28.79 \text{ kN}; \quad \text{s.d.} = 2.09 \text{ kN}$$

Now, applied moment is given by:

$$(P_1 + P_2)\frac{L}{4} + w_s \frac{L^2}{8} = 1.5X_1 + 1.5X_2 + 4.5X_3$$

Hence:

$$Z = X_4 - 1.5X_1 - 1.5X_2 - 4.5X_3$$

or:

$$Z = (y_4\sigma_4 + m_4) - 1.5(y_1\sigma_1 + m_1) - 1.5(y_2\sigma_2 + m_2) - 4.5(y_3\sigma_3 + m_3)$$

The first set of iterations for β will start with the input:

Symbol	Mean	s.d.	Units
X_1	41.47	2.61	kN
X_2	20	5	kN
X_3	5	0.5	kN/m
X_4	1100	100	kNm

which leads to $\beta = 2.75$ after two sets of iterations, and the second set of iterations will have the input:

Symbol	Mean	s.d.	Units
X_1	30	5	kN
X_2	28.79	2.09	kN
X_3	5	0.5	kN/m
X_4	1100	100	kNm

leading, after two sets of iterations, to $\beta = 3.08$.

The reliability index for the system is 2.75. (If the two extreme values are summed together β is equal to 1.92.)

Another popular form of load combination model is that proposed by Ferry Borges and Castanheta (1974) and the reader will find details of this approach in Report 63 of C.I.R.I.A. (1976).

Chapter Nine
The Reliability of Geotechnical Structures

The worked examples throughout this book are intended to illustrate how the level II method of reliability analysis can be applied to civil engineering problems and, because of their introductory nature, are fairly straightforward.

In this final chapter, example 9.2 deals with a practical geotechnical engineering problem which is given to illustrate the application of the level II method and also as a summary of part of the material contained elsewhere in the book.

Geotechnical engineering involves more uncertainties than those encountered in structural engineering and it is therefore necessary to mention some of these problems and briefly describe how they can be dealt with.

There is nothing truly 'random' about the variation of soil properties from one point to another. For example the formation of an estuarine deposit will have been controlled by such agencies as the rate of erosion of the environmental soil, the depth of suspension, the current velocity, etc. although the effect of wind, tidal currents, etc. may well have imposed a random pattern of variation across the overall trend.

However, soil and rock deposits, with properties given them by nature, are amongst the most variable of the materials that a civil engineer is called on to use. Faults can occur quite erratically and tests and measurements carried out are, of necessity, limited, often with the added complication that parameter values from laboratory tests can only approximate the in-situ values.

The standard approach to geotechnical uncertainty is the observational method, evolved by Terzaghi and Peck (1948) and later discussed by Peck (1969) and Casagrande (1965).

Briefly, the technique consists of designing the foundation, retaining wall, etc. using the information available and then, during construction, checking on the original design assumptions by measuring predetermined parameters such as pore water pressures, deformations, etc. In the light of these observations further construction, i.e. rate or dimensions, can be modified.

In the hands of the competent and experienced engineer the observational method can yield satisfactory and economical structures. The method may not be so effective with a relatively inexperienced engineer and, for this reason, factors of safety based on experience gained from similar structures are included in the design although statistics and probability theory provide mathematical approaches that can help to deal with uncertainties and be of assistance with the observational method.

Site investigation procedure

The object of a site investigation for a proposed structure is to determine those properties of the soil that will significantly affect the design and construction. From a statistical point of view this involves obtaining estimates of the means, standard deviations and probability distributions of the relevant parameters.

In order to do this, a major part of a site investigation must involve determining whether different soil layers lie beneath the surface and, if so, the positions of the various soil interfaces. Once the subsoil had been divided into sub-regions then each layer can be regarded as an independent population from which appropriate samples should be collected.

It has been illustrated by Morse (1971) that classifying soils into different sub-regions by visual inspection and field tests may not be sufficiently accurate and can lead to quite wrong design parameters if different, although superficially similar, soils are grouped as belonging to the one population. Obviously, a poor site investigation will result in inadequate subsoil information and the risk of an over-designed structure.

As was confirmed (Butcher, 1984) at a recent British Geotechnical Society's debate, except for a few enlightened cases, the average site investigation is presently carried out in accordance with a set of arbitrary specification clauses, usually prepared with an eye more on costs than on the need to obtain relevant information. If statistical methods are to become widely used in geotechnical design then site investigation procedures will somehow have to be made more flexible. The problem is that the greater the number of representative samples collected and tested the greater the dependability of the reliability analysis prediction but, unfortunately, the greater the cost of the site investigation.

The interpretation of borehole results can often prove difficult and may require considerable experience on the part of the engineer attempting to model the subsoil in this manner. This modelling procedure has little to do with probability theory and, without the right amount of information, the final estimation of the soil profile, although based on accurate information, can be quite wrong. Figure 9.1a shows the information obtained from two boreholes at some distance apart at a particular site and Figs 9.1b, c and d illustrate three possible interpretations. Statistics can be of little help here. What is required is extra information which can only be obtained from another borehole.

For this reason a site investigation should ideally be split into two phases. During the first phase, boreholes should be put down and enough soil samples collected for a laboratory test programme to be

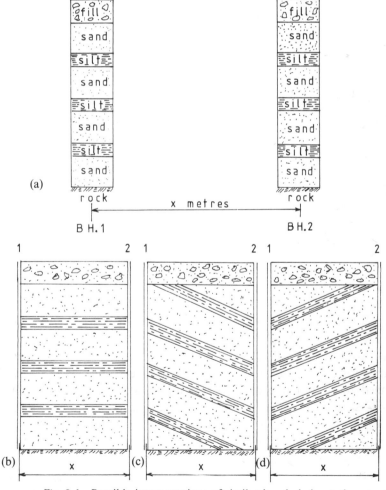

Fig. 9.1 Possible interpretations of similar borehole journals

carried out. With this information assembled and analysed it may become apparent that more data is necessary. In such a situation it would be very useful if the investigation contract somehow allowed for a second phase involving further site investigation work to be carried out.

The probability of the existence of a geological feature cannot be predicted by a simple statistical procedure. Geological forms can only be predicted by some form of inductive reasoning, possibly enhanced by the use of Bayes' theorem, involving any results obtained from the site investigation together with past experience (see example 1.18).

If it is known or suspected that an old slip surface exists and if it is required that its existence, or nonexistence, be proved completely, then its location can only be discovered by continuous coring as part of a systematic search programme.

Soil sampling theory

There is little argument that for quality control, such as the measurement of density along a compacted embankment or the testing of sand from lorries delivering to a site, sample selection should be done on a random basis. However, it has been shown by Lumb (1974), that random selection does not give the best results for in-situ sampling. A systematic pattern of boreholes and the collection of samples at equally spaced vertical intervals for each sub-region is the most efficient method.

Also, it cannot be over-emphasised that, for the test results obtained from a soil sample to be meaningful, the sample must have been taken from the correct soil subregion.

Probabilistic treatment of the substrata

It has been mentioned that a major part of a site investigation is to divide the subsoil into a set of idealised subregions of the different soil types. Only when this procedure has been completed is it possible to carry out meaningful design calculations.

In many cases a straightforward design procedure can be achieved by assuming that the soil layers are locally homogeneous, i.e. they exhibit no significant variations either in thickness or property within the area of the site. However, it should be remembered that on occasions the assumption of homogeneity will be unrealistic. An example is the case of a compressible soil layer whose thickness varies randomly.

Once the various soil horizons have been established, any soil profile characteristic whose uncertainty might have an important effect on the performance predictions can be treated as a random function of either the vertical and/or the horizontal directions (Vanmarke, 1977).

In theory, given sufficient sample measurements, it is possible to predict the value of a soil parameter at any location but the number of test results available will always be limited and the models of soil deposits that are used in reliability analyses are therefore invariably based on one of the following two basic assumptions.

(i) SOIL MASS IS SPATIALLY UNIFORM

A soil mass can be considered to be spatially uniform when it can be assumed that the measured values of a particular parameter vary randomly within it, about the mean value of the measurements.

Throughout the previous chapters spatial uniformity has been tacitly assumed, i.e. the measured values of soil parameters have been assumed to vary randomly with no specific directional trends.

A problem, often encountered with small structures, is when the provisional sum provided for the site investigation precludes any possibility of a comprehensive survey in which directional trends might be recognised. In such a situation the engineer has no choice but to assume spatial uniformity in spite of the errors this may cause.

(i) SOIL MASS HAS DIRECTIONAL TRENDS

There are often occasions when natural soil deposits have parameters that exhibit directional trends. These trends can range in direction from vertical to horizontal and when they are in evidence it is best to allow for them by the use of regression lines. For most soil parameters the best regression line is a straight line.

For a random variable, Y, related to another random variable, X, the standard deviation of Y should be taken as the square root of the variance of Y about the regression line Y on X rather than about the mean value of Y. This value, known as the standard error of the estimate, has already been discussed in chapter 8 and its use is further illustrated in the following example.

Example 9.1

Unconfined compression tests carried out on undisturbed samples taken from a deep clay deposit gave the following results:

Depth (m)	1	2	3	4	5
c_u (kN/m^2)	27	19	34	37	42
		21		40	38
		23		34	

(i) Determine a suitable regression line that will give the relationship between c_u, the strength of the soil, and z, the depth from the surface to the point considered.

(ii) A strip foundation, 2 m wide, is to be founded at a depth of 1.5 m below the surface of the clay. Determine values for the mean and standard deviation of c_u suitable for the evaluation of the probability of bearing capacity failure.

Solution

THE REGRESSION LINE

The equation for the regression line is obtained by the method discussed in chapter 8 or from a suitable pocket calculator.

$$c_u = 5.006z + 15.804 \ kN/m^2$$

where z is the depth (in m) below the ground surface.

THE STANDARD ERROR OF THE ESTIMATE (METHOD i)

The variance of c_u about this line is given by:

$$(s_{cu,z})^2 = \sum \frac{(c_u - c_u(e))^2}{N - 1}$$

where

c_u = measured values of c_u

$c_u(e)$ = expected value of c_u as obtained from the regression line formula

Calculations are best tabulated:

z	c_u	$c_u(e)$	$(c_u - c_u(e))^2$
1	27	20.125	47.266
2	19	25.925	39.627
2	21	25.925	18.447
2	23	25.925	5.267
3	34	30.466	12.489
4	37	35.636	1.861
4	40	35.636	19.045
4	34	35.636	2.677
5	42	40.807	1.423
5	38	40.087	7.879
			$\sum$ 155.981

Therefore:

$$S_{cu,z} = \sqrt{\frac{155.981}{9}} = 4.163$$

THE STANDARD ERROR OF THE ESTIMATE (METHOD ii)
The following values can soon be established:

$$m_z = 3.2; \qquad \sigma_z = 1.3984; \qquad \sum z = 32$$

$$m_{cu} = 31.5; \qquad \sigma_{cu} = 8.3433; \qquad \sum c_u = 315; \qquad \sum zc_u = 1099$$

Now:

$$COV(z, c_u) = \frac{1}{N-1}\left[\sum zc_u - \frac{\sum z . \sum c_u}{N}\right]$$

$$= \tfrac{1}{9}\left[1099 - \frac{32 \times 315}{10}\right] = 10.111$$

and:

$$r = COV(z, c_u) = \frac{10.111}{1.3984 \times 9.3433}$$

$$= 0.8666$$

Therefore:

$$s(c_u, z) = \sigma_{cu}\sqrt{(1 - r^2)} = 8.3433\sqrt{(1 - 0.8666^2)}$$

$$= 4.163$$

DETERMINATION OF VALUES FOR c_u
(a) By the assumption of c_u being purely random:

$$m_{cu} = 31.5 \text{ kN/m}^2; \qquad \sigma c_u = 8.34 \text{ kN/m}^2$$

(b) Using the regression line:
For a foundation supported by a homogeneous soil deposit the values of the soil variables at a depth B below the foundation (where B is the width or diameter of the foundation) are generally taken as representative of the deposit. With this rule $z = 1.5 + 2 = 3.5$ m. Hence:

$$m_{cu} = 5.006 \times 3.5 + 15.804 = 33.05 \text{ kN/m}^2$$

$$\sigma_{cu} = 4.16$$

If we assume that the ultimate bearing pressure is equal to $5.7c_u$ and that the applied bearing pressure will have a mean value of 115 kN/m² and a standard deviation of 15 kN/m², we can compare the effect of choosing each set of values.

Assuming purely random values:

$$\beta = 1.30 \quad (\text{nominal } P_f = 9.68 \times 10^{-2})$$

Allowing for directional trend:

$$\beta = 2.62 \quad (\text{nominal } P_f = 4.40 \times 10^{-3})$$

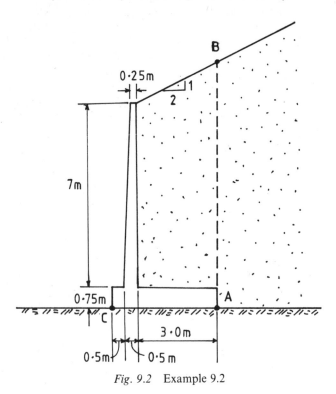

Fig. 9.2   Example 9.2

Example 9.2

Details of a proposed reinforced concrete retaining wall are shown in Fig. 9.2. The fill material will be a granular soil with the following properties: unit weight, γ_1 – mean = 19 kN/m³; s.d. = 1.0 kN/m³; angle of shearing resistance, ϕ_1 – mean = 35°; s.d. = 1.5°.

The soil on which the wall will be founded is a sand/gravel mixture and penetration tests carried out down through the soil showed that the mean value of the unit weight γ_2 of the soil is constant with depth at 19 kN/m³ with a standard deviation of 1.5 kN/m³. These tests also established the following variation of ϕ_2, the angle of shearing resistance of the foundation soil (in degrees), with z, the depth (in m) below the surface of the soil.

z	0.5	1.0	1.5	2.0	2.5	3.0	3.5	4.0	4.5	5.0	5.5	6.0
ϕ_2	36	39	36	40	38	41	37	43	39	43	40	44

Considering bearing capacity effects and assuming that the dimensions and the unit weight of the wall, 24 kN/m³, are constant, determine the factor of safety and the nominal probability of failure.

Assume all relevant variables to have normal distributions.

Solution
DETERMINATION OF THE MEAN AND STANDARD DEVIATION VALUES FOR N_γ
Inserting corresponding N_γ values for the tabulated ϕ values and using the methods of chapter 8, the regression line of N_γ on z can be found:

$$N_\gamma = 16.129z + 31.838 \quad (r = 0.6991)$$

Assuming that the representative value of N_γ will occur at depth B below the foundation of the wall, the mean value of N_γ can be taken to be:

$$m_{N_\gamma} = 16.129 \times 4.0 + 31.838 = 96.35$$

with a standard deviation of:

$$s_{N_\gamma,z} = 41.592\sqrt{(1 - 0.6991^2)} = 29.74$$

The reader might like to check that the regression line of ϕ on z has the equation:

$$\phi = 1.063z° + 36.212°$$

and that:

$$s_{\phi,z} = 1.914°$$

With this information it is simple to show that the representative value of ϕ, for bearing capacity purposes, has a mean value of 40.46° and a standard deviation of 1.91° from which values for the mean and standard deviation of N_γ can be found.

However, slightly more accurate values for the mean and standard deviation are obtained from the the regression line of N_γ on z and this procedure is considered preferable to directly determining values from the representative values of ϕ.

The coefficient of active earth pressure, K_a, can be taken as equal to the Coulomb value (see Smith, 1982):

$$K_a = \left(\frac{\operatorname{cosec} \psi \, \sin(\psi - \phi)}{\sqrt{\sin(\psi + \delta)} + \sqrt{\dfrac{\sin(\phi + \delta)\sin(\phi - i)}{\sin(\psi - i)}}} \right)^2$$

where:

ψ = angle of back of wall to the horizontal
δ = angle of wall friction
i = angle of inclination of surface of retained soil to the horizontal
ϕ = angle of friction of the retained soil.

The mean and standard deviation of K_a can be found by the approximate method described in chapter 5 and have the following values, respectively, 0.4056 and 0.0334.

Height of AB = $7.75 + 1.5 = 9.25$ m
Thrust from soil, $P_a = 0.5 K_a \gamma_1 9.25^2 = 42.78 \gamma_1 K_a$

$$P_{aH} = P_a \cos \phi_1; \qquad P_{aV} = P_a \sin \phi_1$$

R_v = vertical reaction = weight of wall + soil on heel + P_{aV}

$$= 24(0.75 \times 4 + 7 \times 0.375) + \left(7 \times 3 + \frac{3 \times 1.5}{2} \right) \gamma_1$$

$$+ 42.78 K_a \gamma_1 \sin \phi_1$$

For a surface strip footing the ultimate bearing pressure, q_u, is given by the expression:

$$q_u = 0.5 B \gamma N_\gamma i_\gamma$$

where:

B = width of foundation
γ = unit weight of supporting soil
N_γ = bearing capacity coefficient
i_γ = inclined load factor

The ultimate vertical load, Q_u, that can act of the foundation is found from the expression:

$$Q_u = B' q_u$$

where:

$B' = (B - 2e)$
e = eccentricity of R_v

Hence:

$$R = Q_u = 0.5(B - 2e)B\gamma_2 N_\gamma i_\gamma = 2(4 - 2e)\gamma_2 N_\gamma i_\gamma$$

$$S = 135 + 23.25\gamma_1 + 42.78 K_a \gamma_1 \sin \phi_1$$

Therefore:

$$Z = 2(4 - 2e)\gamma_2 N_\gamma i_\gamma - 135 - 23.25\gamma_1 - 42.78 K_a \gamma_1 \sin \phi_1$$

The basic variables are γ_1, γ_2, K_a, $\sin \phi_1$, N_y, i_y and e, and the standard deviations of the first four can be found by the simplified method described in chapter 5. The methods used below for i_y and e were suggested by the author in Research Report No. 46, published by the Transport and Road Research Laboratory (Smith, 1986).

THE INCLINED LOAD FACTOR i_y

Various expressions have been proposed for i_y and the one used in this text is that suggested by Sokolovski (1960):

$$i_y = \left(1 - \frac{P_{aH}}{R_v}\right)^3$$

Now:

$$P_{aH} = \text{horizontal thrust} = 42.78 K_a \gamma_1 \cos \phi_1$$

Therefore:

$$i = \left(1 - \frac{42.78 K_a \gamma_1 \cos \phi_1}{42.78 K_a \gamma_1 \sin \phi_1 + 23.25 \gamma_1 + 135}\right)^3$$

The mean value for i_y (for ϕ_1 equals $35°$ and γ_1 equals 19 kN/m^3) is 0.2713.

For all practical purposes i_y is only sensitive to changes in the value of ϕ and, by assuming that γ_1 is constant at 19 kN/m^3, σ_{i_y} is found (by the approximate method) to be 0.0395.

THE ECCENTRICITY e

$$e = \left| \frac{M}{R_v} - \frac{B}{2} \right| = \left| \frac{M}{R_v} - 2 \right|$$

Taking moments about A, the heel of the wall:

$$\text{Soil pressure} = P_{aH} \times \frac{AB}{3} = \frac{9.25}{3} \times 42.78 K_a \gamma_1 \cos \phi_1$$

$$= 131.91 K_a \gamma_1 \cos \phi_1$$

$$\text{Weight of soil on heel} = \left(7 \times 3 \times 1.5 + \frac{3 \times 1.5 \times 1}{2}\right)\gamma_1$$

$$= 33.75 \gamma_1$$

$$\text{Weight of wall} = 24\left(0.75 \times 4 \times 2 + 0.25 \times 7 \times 3.125\right.$$

$$\left. + \frac{7 \times 0.25 \times 3.333}{2}\right)$$

$$= 345.25 \text{ kN}$$

Hence:

$$e = \left| \frac{345.25 + 33.75\gamma_1 + 131.91 K_a\gamma_1 \cos \phi_1}{135 + 23.25\gamma_1 + 42.78 K_a\gamma_1 \sin \phi_1} - 2 \right|$$

The mean value of e (when ϕ_1 equals $35°$ and γ_1 equals 19 kN/m^3) is 0.3745 m.

By the approximate method, assuming γ_1 constant at 19 kN/m^3, the standard deviation of e works out at 0.0832 m.

Designating the variables as X_1 to X_7:

Variable	Mean	s.d.
$\gamma_1 = X_1$	19	1
$\gamma_2 = X_2$	19	1.5
$K_a = X_3$	0.4056	0.0334
$\sin \phi_1 = X_4$	0.5736	0.0215
$N_\gamma = X_5$	96.35	29.74
$i_\gamma = X_6$	0.2713	0.0395
$e = X_7$	0.3754	0.0832

$$Z = g(X) = 2(4 - 2X_7)X_2 . X_5 . X_6 - 135 - 23.25 X_1$$
$$- 42.78 X_1 . X_3 . X_4$$

$$h(y) = 2[4 - 2(y_7\sigma_7 + m_7)](y_2\sigma_2 + m_2)(y_5\sigma_5 + m_5)(y_6\sigma_6 + m_6)$$
$$- 135 - 23.25(y_1\sigma_1 + m_1)$$
$$- 20.055(y_1\sigma_1 + m_1)(y_3\sigma_3 + m_3)(y_4\sigma_4 + m_4)$$

The iterative procedure for β gives:

Iteration	y_1	y_2	y_3	y_4	y_5	y_6	y_7	β	$h(y)$
1	0.0	0.0	0.0	0.0	0.0	0.0	0.0	0.0	2461
2	0.06	−0.49	0.03	0.01	−1.91	−0.99	0.16	2.17	332
3	0.10	−0.27	0.05	0.02	−2.48	−0.55	0.09	2.56	−93.1
4	0.09	−0.15	0.04	0.02	−2.43	−0.29	0.05	2.46	−10.8
5	0.09	−0.16	0.04	0.02	−2.42	−0.30	0.05	2.45	−0.09
6	0.09	−0.16	0.04	0.02	−2.42	−0.30	0.05	2.45	−0.02

The reliability index is 2.45 and the nominal P_f equals 7×10^{-3}. (A Monte Carlo simulation, after 10 000 iterations, gave β equal to 2.28.)

The factor of safety, based on the mean values of the variables, is equal to 4.21 and it is interesting to consider that, whilst the factor of safety indicates a satisfactory design, the reliability index shows that there is a nominal risk of about 1 in 150.

The reason is of course that the reliability index allows for the variability of the variables, particularly N_y which has a large coefficient of variation and is also a dominant factor in the value of the bearing capacity.

In passing it should be noted that when considering sliding effects, the layer of soil in contact with the foundation of the wall is the surface and as the coefficient of friction is related to the angle of shearing resistance of the soil it is important to use the value of ϕ for z equal to zero, which is obtained from the regression line and is equal to 36.2°.

An example of a fuller reliability analysis of a retaining wall has been prepared by the writer (Smith,1986) and also an illustration of how the method can be applied to propped embedded cantilever walls (Smith, 1985).

Correlated variables in soil mechanics

Although the level II method can easily be adapted to allow for two variables being correlated (see chapter 8) there is little doubt that, for soils problems, the method is simplest to use when the various basic variables can be assumed to be independent.

Except for the three bearing capacity coefficients N_c, N_q and N_y, the assumption of full independence of soil variables is often logical and appears to give reliable results.

Bearing capacity coefficients are functions of the variable ϕ, the angle of shearing resistance of the foundation soil, and are therefore completely correlated with r equal to one. The assumption of independence for these three coefficients can lead to an overestimation of the value of β.

One possible solution, illustrated in the following example, is to determine the linear relationships that N_c and N_y have with N_q. Once these relationships are obtained then the limit state function, Z, can be expressed in terms of the one bearing capacity coefficient, N_q.

Example 9.3

Details of a square reinforced concrete foundation, $3.6 \times 3.6 \text{ m}^2$, are shown in Fig. 9.3.

The foundation is founded at a depth of 5 m below the surface of a

partially saturated silt which has both a high cohesive and a high
frictional strength with the following values:

Cohesion: $m_c = 90 \text{ kN/m}^2$; $\sigma_c = 30 \text{ kN/m}^2$
Angle of friction: $m_\phi = 30°$; $\sigma_\phi = 3°$
Unit weight: $m_\gamma = 19 \text{ kN/m}^3$; $\sigma_\gamma = 0.5 \text{ kN/m}^3$

The foundation will support a concentric 1 m^2 reinforced concrete
column carrying a load of mean value of 25 MN and coefficient of
variation of 10%.
  Determine the reliability index against bearing capacity failure.

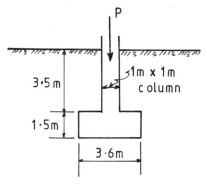

Fig. 9.3 Example 9.3

Note: Soils of this type are notoriously variable in strength and the
drained strength parameters are generally considered to give the best
measure of their strengths (Lumb, 1966).

Solution
Taking the unit weight of concrete as equal to 24 kN/m^3 and assuming
that the unit weight of the excavated soil is constant at 19 kN/m^3:

$$\begin{matrix} \text{weight of} \\ \text{concrete} \end{matrix} - \begin{matrix} \text{weight of} \\ \text{excavated soil} \end{matrix} = (24 - 19)(3.6 \times 1.5^2 + 1 \times 3.5)$$

$$= 114.7 \text{ kN}$$

To obtain linear relationships between N_c, N_γ and N_q we must decide on
a suitable range of values for ϕ, the angle of shearing resistance of the
soil. As ϕ contributes to the resistance component it must reduce during
the iterative procedure to find β and, as its mean value is $30°$ and its
standard deviation is $3°$, a suitable range of ϕ values appears to be from
30 to $30 - 4 \times 3$, i.e. from $30°$ to $18°$.
  The values of the three coefficients for these values of ϕ are given in
appendices V, VI and VII:

ϕ	N_q	N_c	N_γ
18	5.26	13.10	2.08
19	5.80	13.93	2.48
20	6.40	14.83	2.95
21	7.07	15.81	3.50
22	7.82	16.88	4.13
23	8.66	18.05	4.88
24	9.60	19.32	5.75
25	10.66	20.72	6.76
26	11.85	22.25	7.94
27	13.20	23.94	9.32
28	14.72	25.80	10.94
29	16.44	27.86	12.84
30	18.40	30.14	15.07

The linear relationships between N_c and N_q and between N_γ and N_q can be found by the methods described in chapter 8 or with the aid of a suitably programmed calculator:

$$N_c = \quad 6.631 + 1.298N_q$$

$$N_\gamma = -3.472 + 0.985N_q$$

$$Q_u = (1.3cN_c + \gamma z N_q + 0.4\gamma B N_\gamma)B^2 = R$$

Now, B is 3.6 m and z is 5 m. Hence:

$$Q_u = 16.848cN_c + 64.8N_q + 18.662\gamma N_\gamma$$

$$= 16.848(6.631 + 1.298N_q) + 64.8\gamma N_q$$

$$+ 18.662\gamma(-3.472 + 0.985N_q)$$

$$= 117.2c - 64.80\gamma + (21.87c + 83.18\gamma)N_q$$

Now S, the total load on foundation is given by

$$P + 114.7 \text{ kN}$$

where P is the column load. Hence:

$$Z = 117.2c - 64.80\gamma + (21.87c + 83.18\gamma)N_q - P - 114.7$$

There are therefore four basic variables:

Variable	Symbol	Mean (m)	s.d. (σ)	Units
γ	X_1	19	0.5	kN/m^3
c	X_2	90	30	kN/m^2
N_q	X_3	18.4	6.45	
P	X_4	25 000	2500	kN

and:

$$Z = 117.2X_2 - 64.80X_1 + (21.87X_2 + 83.18X_1)X_3 - X_4 - 114.7$$

which can be expressed in reduced variables.

The iterative procedure for β gives:

Iteration	y_1	y_2	y_3	y_4	β	$h(y)$
1	0.0	0.0	0.0	0.0	0.0	49 005.3
2	-0.05	-0.98	-1.46	0.16	1.77	6 091.8
3	-0.03	-0.91	-1.83	0.25	2.06	-471.7
4	-0.02	-0.76	-1.87	0.25	2.04	-148.4
5	-0.02	-0.72	-1.88	0.24	2.03	-9.4
6	-0.02	-0.71	-1.89	0.24	2.03	-0.03

The reliability index equals 2.03. (A check, carried out by Monte Carlo simulation, gave β as 2.17).

Further reading

As was stated in the introduction to this book, the object has been to present a summary of the main aspects of statistics and probability theory that are relevant in civil engineering. The author is all too aware of things that have been left out.

For instance the assumption of spatial uniformity, used throughout the book, is satisfactory for many cases but can lead to misleading results when trend effects are significant as can happen in settlement analyses or slope stability work.

In earth slopes the variability of the soil strength can vary with distance so that, although the value of the mean average strength remains constant, the variance can decrease as the length of the slope increases.

Allowing for this effect makes it possible to predict the most likely length of slope that would be involved in a particular slope failure, (Vanmarke 1977), but much work still remains to be done in this field.

Any reader who wishes to study this subject further is advised to obtain the book by Benjamin and Cornell (1970) and also the book by Thoft-Christensen and Baker (1982).

Bibliography

Alonso, E. E. (1976). 'Risk analysis of slopes and its application to slopes in Canadian sensitive clays', *Geotechnique*, Vol. 26, No. 3.

Benjamin, J. R. and Cornell, C. A. (1970). *Probability, Statistics and Decision for Civil Engineers*, McGraw-Hill Book Co.

Butcher, A. P. (1984). 'The current state of the ground investigation industry in the U.K.', *Ground Engineering*, Vol. 17, No. 4.

Casagrande, A. (1965). 'The role of the calculated risk in earthwork and foundation engineering', *A.S.C.E. S.M.4.*

Construction Industry Research and Information Association (1976). 'Rationalisation of safety and serviceability factors in structural codes', *Report No. 63.*

Cornell, C. A. (1968). 'Engineering seismic risk analysis', *Bulletin of Seismological Soc. of America*, Vol. 58, No. 5.

Cornell, C. A. (1969). *Structural Safety Specifications Based on Second-Moment Reliability Analysis*, I.A.B.S.E. Symposium on Concepts of Safety of Structures and Methods of Design, London.

De Mello, V. F. B. (1977). 'Reflections on design decisions of practical significance to embankment dams', *Geotechnique*, Vol. 27, No. 3.

Der Kieureghian, A. and Ang, A, H.-S. (1977). 'Fault-rupture model for seismic risk analysis', *Bulletin of Seismologial Soc. of America*, Vol. 67, No. 4.

Ditlevsen, O. (1973). 'Structural reliability and the invariance problem', Solid Mechanics Division, University of Waterloo (Ontario), *Research Report*, 22. 1973.

Ditlevsen, O. (1976). *Evaluation of the effect on structural reliability of slight deviations from hyperplane limit state surfaces*, D.I.A.L.O.G. 2–76, Danmarks Ingeniorakademi, Lyngby.

Ditlevsen, O. (1982). *Basic Reliability Concepts*, N.A.T.O. Advanced Study Institute on reliability theory and its application in structural and soil mechanics, Bornholm, Denmark.

Ferry Borges, J. and Castanheta, M. (1974). *Structural Safety*. National Laboratory of Civil Engineering (Lisbon).

Fiessler, B. (1980). *Das programmsystem FORM zur berechnung der versagenswahrscheinlichkeit vo komponenten von tragsystmen*, Sonderforschungsbereich 96, Technical University of Munich.

Fiessler, B., Neumann, H.-J. and Rackwitz, R. (1979). 'Quadratic limit states in structural reliability', *J. Eng. Mech. Div.*, *A.S.C.E.*, Vol. 105.

Fisher, R. A. and Tippett, L. H. C. (1928). 'Limiting forms of the frequency distribution of the largest or smallest member of a sample', *Proc. Cambridge Philosophical Society*, Vol. 24.

Grigoriu, M. (1983). 'Risk analysis for code calibration', *Structural Safety*, Vol. 1, No. 4, Elsevier Science Publishers B.V., Amsterdam.

Gumbel, E. J. (1958). *Statistics of Extremes*, Columbia University Press.

Hansen, J. B. (1970). 'A revised and extended formula for bearing capacity', *Institut Kobenhavn Bulletin*.

Hasofer, A. M. and Lind, N. C. (1974). 'Exact and invariant second-moment code format', *J. Eng. Mechs. Div.*, *A.S.C.E.*, Vol. 100.

Hooper, J. A. and Butler, F. G. (1966). 'Some numerical results concerning the shear strength of London clay', *Geotechnique*, Vol. 16, No. 4.

Jennings, A. (1977). '*Matrix Computation for Engineers and Scientists*', John Wiley and Sons, Chichester.

Lumb, P. (1965). 'The residual soils of Hong Kong', *Geotechnique*, Vol. 15, No. 1.

Lumb, P. (1966). 'The variability of natural soils', *Canadian Geotechnical J.*, Vol. 3, No. 2.

Lumb, P. (1970). 'Safety factors and the probability distribution of soil strength', *Canadian Geotechnical J.*, Vol. 7, No. 3.

Lumb, P. (1974). 'Applications of statistics in soil mechanics', *Soil mechanics – New horizons*, Butterworth, London.

Mayer, H. (1926). *Die Sicherheit der Bauwerke*, Springer-Verlag, Berlin.

Meyerhof, G. G. (1955). 'Influence of roughness of base and groundwater conditions on the ultimate bearing capacity of foundations', *Geotechnique*, Vol. 5, No. 2.

Meyerhof, G. G. (1970). 'Safety factors in soil mechanics', *Canadian Geotechnical J.* Vol. 7, No. 4.

Meyerhof, G. G. (1982). 'Limit state designs in geotechnical engineering', *Structural Safety*, Vol. 1, No. 1, Elsevier Science Publishers B.V., Amsterdam.

Morse, R. K. (1971). 'The Importance of proper soil units for statistical analysis', *Statistics and Probability in Civil Engineering*, ed. P. Lumb, Hong Kong University Press, Hong Kong.

Mostyn, G. (1983). *A Statistical Approach to Characterising the Permea-*

bility of a Mass, Fourth Int. Conf. on Application of Statistics and Probability in Soil and Structural Engineering.

Neave, H. R. (1978). *Statistics Tables for Mathematicians, Engineers, Economists and the Behavioural and Management Sciences*, George Allen & Unwin Ltd, London.

Peck, R. B. (1969). 'Advantages and limitations of the observational method in applied soil mechanics', *Geotechnique*, Vol. 19, No. 2.

Rackwitz, R. (1976). *Practical probabilistic approach to design*, technical University of Munich.

Rackwitz, R. and Fiessler, B. (1978). 'Structural reliability under combined random load sequences', *Compt. Struct., Vol. 9.*

Ravindra, M. K., Heaney, A. C. and Lind, N. C. (1969). *Probabilistic evaluation of safety factors*, I.A.B.S.E. Symposium on Concepts of Safety of Structures and Methods of Design, London.

Reynolds, C. E. (1957). *Reinforced Concrete Designers' Handbook*, Concrete Publications Ltd, 5th ed., London.

Rosenblueth, E. and Esteva, L. (1972). 'Reliability basis for some Mexican codes', *American Concrete Institute Publication SP-31.*

Saetre, H. J. (1975). *On High Wave Conditions in the Northern North Sea*, Oceanology International 75, Conference papers, Brighton, U.K.

Schultze, E. (1972). *Frequency Distributions and Correlations of Soil Properties*, First Int. Conf. on Applications of Statistics and Probability in Soil and Structural Engineering.

Smith, G. N. (1982). *Elements of Soil Mechanics for Civil and Mining Engineers*, 5th ed., Granada Ltd, London.

Smith, G. N. (1985). 'The use of probability theory to assess the safety of propped embedded cantilever retaining walls', *Geotechnique*, Vol. 35, No. 4.

Smith, G. N. (1986). 'A suggested method for reliability analysis of earth retaining structures', *Research Report No. 46*, Transport and Road Research Laboratory, Crowthorne.

Sokolovski, V. V. (1960). *Statistics of Soil Media*, Butterworth, London.

Tang, W. H. (1981). 'Probabilistic evaluation of loads', *J. Eng. Mech. Div., A.S.C.E.*, Vol. 107.

Terzaghi, K. and Peck, R. B. (1948). *Soil Mechanics in Engineering Practice*, J. Wiley & Son, New York.

Thoft-Christensen, P. and Baker, M. J. (1982). *Strutural Reliability Theory and its Applications*, Springer-Verlag, Berlin, Heidelberg, New York.

Turkstra, C. J. (1970). 'Theory of structural design decisions', *Solid Mechanics Study 2*, University of Waterloo (Ontario).

Turnbull, W. J., Compton, J. R. and Ahlvin, R. G. (1966). 'Quality control of compacted earthwork', *J. Soil Mechs. and Found. Div., A.S.C.E.*, Vol. 92.

Uno, T., Arai, H. and Shibayama, M. (1981). *Models for Predicting Ground Water Level*, I.C.S.M.F.E., Stockholm.

Vanmarke, E. (1977). 'Probabilistic modelling of soil profiles', *J. Geotechnical Eng. Div., A.S.C.E.*, Vol. 103.

Vanmarke, E. (1977). 'Reliability of earth slopes', *J. Geotechnical Eng. Div., A.S.C.E.*, Vol. 103.

Yuceman, M. S. and Tang, W. H. (1975). 'Long Term Stability of Slopes – a Reliability Approach', Second Int. Conf. on Application of Statistics and Probability in Soil and Structural Engineering.

Appendix I

Simpson's Rule for Approximate Integration

An approximate method for determining the area under a curve and between two abscissae, x_1 and x_n, is known as Simpson's rule and can often prove useful in statistics.

The procedure is to divide the distance $x_n - x_1$ into an *even* number of equal spaces, of dimension a. The abcissae $x_2, x_3, x_4, x_5, \ldots, x_{n-1}$ are therefore obtained.

If the ordinates $y_1, y_2, y_3, y_4, \ldots, y_n$, corresponding to $x_1, x_2, x_3, x_4, \ldots, x_n$ are determined, then an approximate value, A, for the area under the curve can be obtained:

$$A = \frac{a}{3} [(y_1 + y_n) + 4(y_2 + y_4 + y_6 + \cdots + y_{n-1})$$

$$+ 2(y_3 + y_5 + y_7 + \cdots + y_{n-2})]$$

Example

Determine the area beneath the curve $y = 4 - 0.25x^2$ between the limits $x = -4$ to $x = +4$.

Solution
(a) By integration

$$A = \int_{-4}^{4} [4 - 0.25x^2] \, dx = \left[4x - \frac{0.25x^3}{3} \right]_{-4}^{4} = 21.34$$

(b) By Simpson's rule
Range of x values is from -4 to $+4$ and, if we take a as 1.0, we obtain an even number of spaces (8) and Simpson's rule can be applied.

x	-4	-3	-2	-1	0	1	2	3	4
y	0	1.75	3.0	3.75	4.0	3.75	3.0	1.75	0

Hence:

$$A = \tfrac{1}{3}[(0 + 0) + 4(1.75 + 3.75 + 3.75 + 1.75)$$
$$+ 2(3.0 + 4.0 + 3.0)]$$
$$= 21.33$$

Appendix II

Ordinates of the Standard Normal Curve — $f_Y(y)$

y	0	1	2	3	4	5	6	7	8	9
0.0	0.3989	0.3989	0.3989	0.3988	0.3986	0.3984	0.3982	0.3989	0.3977	0.3973
0.1	0.3970	0.3965	0.3961	0.3956	0.3951	0.3945	0.3939	0.3932	0.3925	0.3918
0.2	0.3910	0.3902	0.3894	0.3885	0.3876	0.3867	0.3857	0.3847	0.3836	0.3825
0.3	0.3814	0.3802	0.3790	0.3778	0.3765	0.3752	0.3739	0.3725	0.3712	0.3697
0.4	0.3683	0.3668	0.3653	0.3637	0.3621	0.3605	0.3589	0.3572	0.3555	0.3538
0.5	0.3521	0.3503	0.3485	0.3467	0.3448	0.3429	0.3410	0.3391	0.3372	0.3352
0.6	0.3332	0.3312	0.3292	0.3271	0.3251	0.3230	0.3209	0.3187	0.3166	0.3144
0.7	0.3123	0.3101	0.3079	0.3056	0.3034	0.3011	0.2989	0.2966	0.2943	0.2929
0.8	0.2987	0.2874	0.2850	0.2827	0.2803	0.2780	0.2756	0.2732	0.2709	0.2685
0.9	0.2661	0.2637	0.2613	0.2589	0.2565	0.2541	0.2516	0.2492	0.2468	0.2444
1.0	0.2420	0.2396	0.2371	0.2347	0.2323	0.2299	0.2275	0.2251	0.2227	0.2203
1.1	0.2179	0.2155	0.2131	0.2107	0.2083	0.2059	0.2036	0.2012	0.1989	0.1965
1.2	0.1942	0.1919	0.1895	0.1872	0.1849	0.1826	0.1804	0.1781	0.1758	0.1736
1.3	0.1714	0.1691	0.1669	0.1647	0.1626	0.1604	0.1582	0.1561	0.1539	0.1518
1.4	0.1497	0.1476	0.1456	0.1435	0.1415	0.1394	0.1374	0.1354	0.1334	0.1315
1.5	0.1295	0.1276	0.1257	0.1238	0.1219	0.1200	0.1182	0.1163	0.1145	0.1127
1.6	0.1109	0.1092	0.1074	0.1057	0.1040	0.1023	0.1006	0.0989	0.0973	0.0957
1.7	0.0940	0.0925	0.0909	0.0893	0.0878	0.0863	0.0848	0.0833	0.0818	0.0804
1.8	0.0790	0.0775	0.0761	0.0748	0.0734	0.0721	0.0707	0.0694	0.0681	0.0669
1.9	0.0656	0.0644	0.0632	0.0620	0.0608	0.0596	0.0584	0.0573	0.0562	0.0551
2.0	0.0540	0.0529	0.0519	0.0508	0.0498	0.0488	0.0478	0.0468	0.0459	0.0449
2.1	0.0440	0.0431	0.0422	0.0413	0.0404	0.0396	0.0387	0.0379	0.0371	0.0363
2.2	0.0355	0.0347	0.0339	0.0332	0.0325	0.0317	0.0310	0.0303	0.0297	0.0290
2.3	0.0283	0.0277	0.0270	0.0264	0.0258	0.0252	0.0246	0.0241	0.0235	0.0229
2.4	0.0224	0.0219	0.0213	0.0208	0.0203	0.0198	0.0194	0.0189	0.0184	0.0180
2.5	0.0175	0.0171	0.0167	0.0163	0.0158	0.0154	0.0151	0.0147	0.0143	0.0139
2.6	0.0136	0.0132	0.0129	0.0126	0.0122	0.0119	0.0116	0.0113	0.0110	0.0107
2.7	0.0104	0.0101	0.0099	0.0096	0.0093	0.0091	0.0088	0.0086	0.0084	0.0081
2.8	0.0079	0.0077	0.0075	0.0073	0.0071	0.0069	0.0067	0.0065	0.0063	0.0061
2.9	0.0060	0.0058	0.0056	0.0055	0.0053	0.0051	0.0050	0.0048	0.0047	0.0046
3.0	0.0044	0.0043	0.0042	0.0040	0.0039	0.0038	0.0037	0.0036	0.0035	0.0034
3.1	0.0033	0.0032	0.0031	0.0030	0.0029	0.0028	0.0027	0.0026	0.0025	0.0025
3.2	0.0024	0.0023	0.0022	0.0022	0.0021	0.0020	0.0020	0.0019	0.0018	0.0018
3.3	0.0017	0.0017	0.0016	0.0016	0.0015	0.0015	0.0014	0.0014	0.0013	0.0013
3.4	0.0012	0.0012	0.0012	0.0011	0.0011	0.0010	0.0010	0.0010	0.0009	0.0009
3.5	0.0009	0.0008	0.0008	0.0008	0.0008	0.0007	0.0007	0.0007	0.0007	0.0006
3.6	0.0006	0.0006	0.0006	0.0005	0.0005	0.0005	0.0005	0.0005	0.0005	0.0004
3.7	0.0004	0.0004	0.0004	0.0004	0.0004	0.0004	0.0003	0.0003	0.0003	0.0003
3.8	0.0003	0.0003	0.0003	0.0003	0.0003	0.002	0.0002	0.0002	0.0002	0.0002
3.9	0.0002	0.0002	0.0002	0.0002	0.0002	0.0002	0.0002	0.0002	0.0001	0.0001

Appendix III
The Cumulative Normal Distribution Function — $F_Y(y)$ or $\Phi(y)$

y	0	1	2	3	4	5	6	7	8	9
−3.9	0.0000	0.0000	0.0000	0.0000	0.0000	0.0000	0.0000	0.0000	0.0000	0.0000
−3.8	0.0001	0.0001	0.0001	0.0001	0.0001	0.0001	0.0001	0.0001	0.0001	0.0001
−3.7	0.0001	0.0001	0.0001	0.0001	0.0001	0.0001	0.0001	0.0001	0.0001	0.0001
−3.6	0.0002	0.0002	0.0001	0.0001	0.0001	0.0001	0.0001	0.0001	0.0001	0.0001
−3.5	0.0002	0.0002	0.0002	0.0002	0.0002	0.0002	0.0002	0.0002	0.0002	0.0002
−3.4	0.0003	0.0003	0.0003	0.0003	0.0003	0.0003	0.0003	0.0003	0.0003	0.0002
−3.3	0.0005	0.0005	0.0005	0.0004	0.0004	0.0004	0.0004	0.0004	0.0004	0.0003
−3.2	0.0007	0.0007	0.0006	0.0006	0.0006	0.0006	0.0006	0.0005	0.0005	0.0005
−3.1	0.0010	0.0009	0.0009	0.0009	0.0008	0.0008	0.0008	0.0008	0.0007	0.0007
−3.0	0.0013	0.0013	0.0013	0.0012	0.0012	0.0011	0.0011	0.0011	0.0010	0.0010
−2.9	0.0019	0.0018	0.0017	0.0016	0.0016	0.0015	0.0015	0.0015	0.0014	0.0014
−2.8	0.0026	0.0025	0.0024	0.0023	0.0023	0.0022	0.0021	0.0021	0.0020	0.0019
−2.7	0.0035	0.0034	0.0033	0.0032	0.0031	0.0030	0.0029	0.0028	0.0027	0.0026
−2.6	0.0047	0.0045	0.0044	0.0043	0.0041	0.0040	0.0039	0.0038	0.0037	0.0036
−2.5	0.0062	0.0060	0.0059	0.0057	0.0055	0.0054	0.0052	0.0051	0.0049	0.0048
−2.4	0.0082	0.0080	0.0078	0.0075	0.0073	0.0071	0.0069	0.0068	0.0066	0.0064
−2.3	0.0107	0.0104	0.0102	0.0099	0.0096	0.0094	0.0091	0.0089	0.0087	0.0084
−2.2	0.0139	0.0136	0.0132	0.0129	0.0125	0.0122	0.0119	0.0116	0.0113	0.0110
−2.1	0.0179	0.0174	0.0170	0.0166	0.0162	0.0158	0.0154	0.0150	0.0146	0.0143
−2.0	0.0228	0.0222	0.0217	0.0212	0.0207	0.0202	0.0197	0.0192	0.0188	0.0183
−1.9	0.0287	0.0281	0.0274	0.0268	0.0262	0.0256	0.0250	0.0244	0.0239	0.0233
−1.8	0.0359	0.0351	0.0344	0.0336	0.0329	0.0322	0.0314	0.0307	0.0301	0.0294
−1.7	0.0446	0.0436	0.0427	0.0418	0.0409	0.0401	0.0392	0.0384	0.0375	0.0367
−1.6	0.0548	0.0537	0.0526	0.0516	0.0505	0.0495	0.0485	0.0475	0.0465	0.0455
−1.5	0.0668	0.0655	0.0643	0.0630	0.0618	0.0606	0.0594	0.0582	0.0571	0.0559
−1.4	0.0808	0.0793	0.0778	0.0764	0.0749	0.0735	0.0721	0.0708	0.0694	0.0681
−1.3	0.0968	0.0951	0.0934	0.0918	0.0901	0.0885	0.0869	0.0853	0.0838	0.0823
−1.2	0.1151	0.1131	0.1112	0.1093	0.1075	0.1056	0.1038	0.1020	0.1003	0.0985
−1.1	0.1357	0.1335	0.1314	0.1292	0.1271	0.1251	0.1230	0.1210	0.1190	0.1170
−1.0	0.1587	0.1562	0.1539	0.1515	0.1492	0.1469	0.1446	0.1423	0.1401	0.1379
−0.9	0.1841	0.1814	0.1788	0.1762	0.1736	0.1711	0.1685	0.1660	0.1635	0.1611
−0.8	0.2119	0.2090	0.2061	0.2033	0.2005	0.1977	0.1949	0.1922	0.1894	0.1867
−0.7	0.2420	0.2389	0.2358	0.2327	0.2296	0.2266	0.2236	0.2206	0.2177	0.2148
−0.6	0.2743	0.2709	0.2676	0.2643	0.2611	0.2578	0.2546	0.2514	0.2483	0.2451
−0.5	0.3085	0.3050	0.3015	0.2981	0.2946	0.2912	0.2877	0.2843	0.2810	0.2776
−0.4	0.3446	0.3409	0.3372	0.3336	0.3300	0.3264	0.3228	0.3192	0.3156	0.3121
−0.3	0.3821	0.3783	0.3745	0.3707	0.3669	0.3632	0.3594	0.3557	0.3520	0.3483
−0.2	0.4207	0.4168	0.4129	0.4090	0.4052	0.4013	0.3974	0.3936	0.3897	0.3859
−0.1	0.4602	0.4562	0.4522	0.4483	0.4443	0.4404	0.4364	0.4325	0.4286	0.4247
−0.0	0.5000	0.4960	0.4920	0.4880	0.4840	0.4801	0.4761	0.4721	0.4681	0.4641

y	0	1	2	3	4	5	6	7	8	9
0.0	0.5000	0.5040	0.5080	0.5120	0.5160	0.5199	0.5239	0.5279	0.5319	0.5395
0.1	0.5398	0.5438	0.5478	0.5517	0.5557	0.5596	0.5636	0.5675	0.5714	0.5753
0.2	0.5793	0.5832	0.5871	0.5910	0.5948	0.5987	0.6026	0.6064	0.6103	0.6141
0.3	0.6179	0.6217	0.6255	0.6293	0.6331	0.6368	0.6406	0.6443	0.6480	0.6517
0.4	0.6554	0.6591	0.6628	0.6664	0.6700	0.6736	0.6772	0.6808	0.6844	0.6879
0.5	0.6915	0.6950	0.6985	0.7019	0.7054	0.7088	0.7123	0.7157	0.7190	0.7224
0.6	0.7257	0.7291	0.7324	0.7357	0.7389	0.7422	0.7454	0.7486	0.7517	0.7549
0.7	0.7580	0.7611	0.7642	0.7673	0.7704	0.7734	0.7764	0.7794	0.7823	0.7852
0.8	0.7881	0.7910	0.7939	0.7967	0.7995	0.8023	0.8051	0.8078	0.8106	0.8133
0.9	0.8159	0.8186	0.8212	0.8238	0.8264	0.8289	0.8315	0.8340	0.8365	0.8389
1.0	0.8413	0.8438	0.8461	0.8485	0.8508	0.8531	0.8554	0.8577	0.8599	0.8621
1.1	0.8643	0.8865	0.8686	0.8708	0.8729	0.8749	0.8770	0.8790	0.8810	0.8830
1.2	0.8849	0.8869	0.8888	0.8907	0.8925	0.8944	0.8962	0.8980	0.8997	0.9015
1.3	0.9032	0.9049	0.9066	0.9082	0.9099	0.9115	0.9131	0.9147	0.9162	0.9177
1.4	0.9192	0.9207	0.9222	0.9236	0.9251	0.9265	0.9279	0.9292	0.9306	0.9319
1.5	0.9332	0.9345	0.9357	0.9370	0.9382	0.9394	0.9406	0.9418	0.9429	0.9441
1.6	0.9452	0.9463	0.9474	0.9484	0.9495	0.9505	0.9515	0.9525	0.9535	0.9545
1.7	0.9554	0.9564	0.9573	0.9582	0.9591	0.9599	0.9608	0.9616	0.9625	0.9633
1.8	0.9641	0.9649	0.9656	0.9664	0.9671	0.9678	0.9686	0.9693	0.9699	0.9706
1.9	0.9713	0.9719	0.9726	0.9732	0.9738	0.9744	0.9750	0.9756	0.9761	0.9767
2.0	0.9773	0.9778	0.9783	0.9788	0.9793	0.9798	0.9803	0.9808	0.9812	0.9817
2.1	0.9821	0.9826	0.9830	0.9834	0.9838	0.9842	0.9846	0.9850	0.9854	0.9857
2.2	0.9861	0.9865	0.9868	0.9871	0.9875	0.9878	0.9881	0.9884	0.9887	0.9890
2.3	0.9893	0.9896	0.9898	0.9901	0.9904	0.9906	0.9909	0.9911	0.9913	0.9916
2.4	0.9918	0.9920	0.9922	0.9925	0.9927	0.9929	0.9931	0.9932	0.9934	0.9936
2.5	0.9938	0.9940	0.9941	0.9943	0.9945	0.9946	0.9948	0.9949	0.9951	0.9952
2.6	0.9953	0.9955	0.9956	0.9957	0.9959	0.9960	0.9961	0.9962	0.9963	0.9964
2.7	0.9965	0.9966	0.9967	0.9968	0.9969	0.9970	0.9971	0.9972	0.9973	0.9974
2.8	0.9974	0.9975	0.9976	0.9977	0.9977	0.9978	0.9979	0.9980	0.9980	0.9981
2.9	0.9981	0.9982	0.9983	0.9983	0.9984	0.9984	0.9985	0.9985	0.9986	0.9986
3.0	0.9987	0.9987	0.9987	0.9988	0.9988	0.9989	0.9989	0.9989	0.9990	0.9990
3.1	0.9990	0.9991	0.9991	0.9991	0.9992	0.9992	0.9992	0.9992	0.9993	0.9993
3.2	0.9993	0.9993	0.9994	0.9994	0.9994	0.9994	0.9994	0.9995	0.9995	0.9995
3.3	0.9995	0.9995	0.9996	0.9996	0.9996	0.9996	0.9996	0.9996	0.9996	0.9992
3.4	0.9997	0.9997	0.9997	0.9997	0.9997	0.9998	0.9998	0.9998	0.9998	0.9998
3.5	0.9998									
3.6	0.9998									
3.7	0.9999									
3.8	0.9999									
3.9	0.9999									

Appendix IV

Ordinates of the Cumulative Student t Distribution Function – $F_T(t)$

v \ P	0.75	0.90	0.95	0.975	0.99	0.995	0.999
1	1.00	3.08	6.31	12.71	31.82	63.66	318.31
2	0.82	1.89	2.92	4.30	6.97	9.93	22.33
3	0.77	1.64	2.35	3.18	4.54	5.84	10.21
4	0.74	1.53	2.13	2.78	3.75	4.60	7.17
5	0.73	1.48	2.02	2.57	3.37	4.03	5.89
6	0.72	1.44	1.94	2.45	3.14	3.71	5.21
7	0.71	1.42	1.90	2.37	3.00	3.50	4.79
8	0.71	1.40	1.86	2.31	2.90	3.36	4.50
9	0.70	1.38	1.83	2.26	2.82	3.26	4.30
10	0.70	1.37	1.81	2.29	2.76	3.17	4.14
11	0.70	1.36	1.80	2.20	2.72	3.11	4.03
12	0.70	1.36	1.78	2.18	2.68	3.06	3.93
13	0.69	1.35	1.77	2.16	2.65	3.01	3.85
14	0.69	1.35	1.76	2.15	2.62	2.98	3.79
15	0.69	1.34	1.75	2.13	2.60	2.95	3.73
16	0.69	1.34	1.75	2.12	2.58	2.92	3.69
17	0.69	1.33	1.74	2.11	2.57	2.90	3.65
18	0.69	1.33	1.73	2.10	2.55	2.89	3.61
19	0.69	1.33	1.73	2.09	2.54	2.86	3.58
20	0.69	1.33	1.73	2.09	2.53	2.85	3.55
21	0.69	1.32	1.72	2.08	2.52	2.83	3.53
22	0.69	1.32	1.72	2.07	2.51	2.82	3.51
23	0.69	1.32	1.71	2.07	2.50	2.81	3.49
24	0.69	1.32	1.71	2.06	2.49	2.80	3.47
25	0.68	1.32	1.71	2.06	2.49	2.79	3.45
26	0.68	1.32	1.71	2.06	2.48	2.78	3.44
27	0.68	1.31	1.70	2.05	2.47	2.77	3.42
28	0.68	1.31	1.70	2.05	2.47	2.76	3.41
29	0.68	1.31	1.70	2.05	2.46	2.76	3.40
30	0.68	1.31	1.70	2.04	2.46	2.75	3.39

v \ P	0.75	0.90	0.95	0.975	0.99	0.995	0.999
40	0.68	1.30	1.68	2.02	2.42	2.70	3.31
50	0.68	1.30	1.68	2.01	2.40	2.68	3.26
60	0.68	1.30	1.67	2.00	2.39	2.66	3.23
70	0.68	1.29	1.67	1.99	2.38	2.65	3.21
80	0.68	1.29	1.66	1.99	2.37	2.64	3.20
90	0.68	1.29	1.66	1.99	2.37	2.63	3.18
100	0.68	1.29	1.66	1.98	2.36	2.63	3.17
150	0.68	1.29	1.66	1.98	2.35	2.61	3.16
∞	0.67	1.28	1.65	1.96	2.33	2.58	3.09

Appendix V

Numerical Values of N_c and its First Derivative

ϕ (degrees)	N_c	$\partial N_c/\partial \phi$
0.00	5.14	12.80
1.00	5.38	14.03
2.00	5.63	14.90
3.00	5.90	15.84
4.00	6.19	16.86
5.00	6.49	17.96
6.00	6.81	19.16
700	7.16	20.46
8.00	7.53	21.87
9.00	7.92	23.40
10.00	8.34	25.07
11.00	8.80	26.89
12.00	9.28	28.88
13.00	9.81	31.06
14.00	10.37	33.45
15.00	10.98	36.07
16.00	11.63	38.96
17.00	12.34	42.14
18.00	13.10	45.64
19.00	13.93	49.52
20.00	14.83	53.82
21.00	15.81	58.59
22.00	16.88	63.89
23.00	18.05	69.80
24.00	19.32	76.41
25.00	20.72	83.81
26.00	22.25	92.12
27.00	23.94	101.47
28.00	25.80	112.02
29.00	27.86	123.96
30.00	30.14	137.50
31.00	32.67	152.92
32.00	35.49	170.52
33.00	38.64	190.68
34.00	42.16	213.85
35.00	46.12	240.56
36.00	50.59	271.49
37.00	55.63	307.43

ϕ (degrees)	N_c	$\partial N_c / \partial \phi$
38.00	61.35	349.37
39.00	67.87	398.51
40.00	75.31	456.36
41.00	83.86	524.78
42.00	93.71	606.11
43.00	105.11	703.28
44.00	118.37	820.04
45.00	133.87	961.17
46.00	152.10	1132.81
47.00	173.64	1342.94
48.00	199.26	1602.00
49.00	229.92	1923.77
50.00	266.88	2326.62

Appendix VI

Numerical Values of N_q and its First Derivative

ϕ (degrees)	N_q	$\partial N_q / \partial \phi$
0.00	1.00	5.14
1.00	1.09	5.63
2.00	1.20	6.16
3.00	1.31	6.75
4.00	1.43	7.39
5.00	1.57	8.11
6.00	1.72	8.90
7.00	1.88	9.78
8.00	2.06	10.75
9.00	2.25	11.83
10.00	2.47	13.02
11.00	2.71	14.36
12.00	2.97	15.84
13.00	3.26	17.50
14.00	3.59	19.36
15.00	3.94	21.43
16.00	4.34	23.76
17.00	4.77	26.37
18.00	5.26	29.32
19.00	5.80	32.64
20.00	6.40	36.39
21.00	7.07	40.63
22.00	7.82	45.45
23.00	8.66	50.93
24.00	9.60	57.17
25.00	10.66	64.31
26.00	11.85	72.48
27.00	13.20	81.86
28.00	14.72	92.66
29.00	16.44	105.13
30.00	18.40	119.57
31.00	20.63	136.35
32.00	23.18	155.90
33.00	26.09	178.76
34.00	29.44	205.59
35.00	33.30	237.18
36.00	37.75	274.54
37.00	42.92	318.89

ϕ (degrees)	N_q	$\partial N_q/\partial\phi$
38.00	48.93	371.76
39.00	55.96	435.08
40.00	64.20	511.27
41.00	73.90	603.41
42.00	85.37	715.42
43.00	99.01	852.33
44.00	115.13	1020.66
45.00	134.87	1228.92
46.00	158.50	1488.25
47.00	187.21	1813.45
48.00	222.30	2224.23
49.00	265.50	2747.24
50.00	319.06	3418.69

Appendix VII
Numerical Values of N_γ and its First Derivative

ϕ (degrees)	N_γ	$\partial N_\gamma/\partial\phi$
0.00	0.00	0.00
1.00	0.00	0.29
2.00	0.01	0.62
3.00	0.02	1.00
4,00	0.05	1.43
5.00	0.07	1.92
6.00	0.11	2.49
7.00	0.16	3.14
8.00	0.22	3.88
9.00	0.30	4.74
10.00	0.39	5.72
11.00	0.50	6.85
12.00	0.63	8.15
13.00	0.78	9.64
14.00	0.97	11.36
15.00	1.18	13.34
16.00	1.43	15.63
17.00	1.73	18.28
18.00	2.08	21.35
19.00	2.48	24.91
20.00	2.95	29.04
21.00	3.50	33.85
22.00	4.13	39.45
23.00	4.88	45.99
24.00	5.75	53.65
25.00	6.76	62.63
26.00	7.94	73.18
27.00	9.32	85.61
28.00	10.94	100.30
29.00	12.84	117.70
30.00	15.07	138.36
31.00	17.69	162.97
32.00	20.79	192.38
33.00	24.44	227.65
34.00	28.77	270.07
35.00	33.92	321.31
36.00	40.05	383.43
37.00	47.38	459.03

ϕ (degrees)	N_γ	$\partial N_\gamma / \partial \phi$
38.00	56.17	551.46
39.00	66.76	664.98
40.00	79.54	805.05
41.00	95.05	978.78
42.00	113.96	1195.41
43.00	137.10	1467.08
44.00	165.58	1809.82
45.00	200.81	2245.00
46.00	244.65	2801.29
47.00	299.52	3517.53
48.00	368.67	4446.79
49.00	456.40	5662.28
50.00	568.57	7266.03

Index

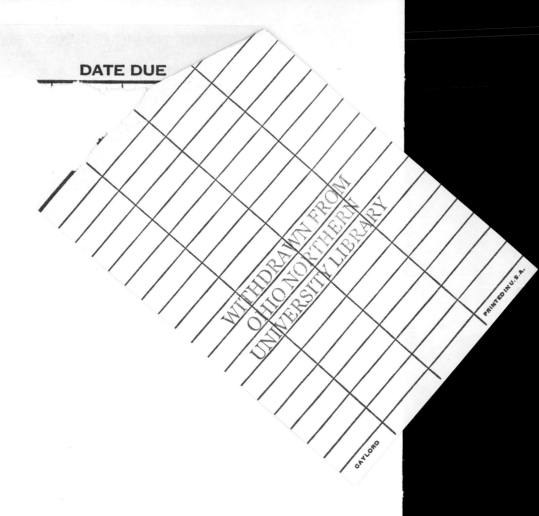